# New Perspectives on Aloe

# New Perspectives on Aloe

Young In Park
Seung Ki Lee

Editors

With 85 Figures, Including 16 Color Plates

 Springer

Young In Park
Korea University
Seoul, Korea
yipark@korea.ac.kr

Seung Ki Lee
Seoul National University
Seoul, Korea
sklcrs@snu.ac.kr

Library of Congress Control Number: 2005939017

ISBN-10: 0-387-31799-6
ISBN-13: 978-0387-31799-1

Printed on acid-free paper.

Printed in the United States of America.     (EB)

9 8 7 6 5 4 3 2 1

springer.com

# Preface

The CAP project for comprehensive aloe research was launched in June 1993 and completed its first stage in 2001. I am deeply moved by this publication containing eight years' research outcome on aloe.

Before launching the CAP project, some members of the CAP research team did not know the real nature of aloe, or simply recognized it as a "functional health food or a dietary supplements" that is used as more of a panacea. In January 1993, I was invited to visit Aloecorp, an America-based local corporation and aloe plantation located in Harlingen, Texas and Tampico, Mexico. Together with several members of CAP research, I took part in that journey, and all of us were very impressed with the aloe industry. In particular, the seminar organized by American aloe researchers was so impressive it led us to have special interest in aloe. In retrospect, we maintained continuous discussions on aloe around the clock. We felt that we became very enthusiastic about aloe, ended up calling ourselves "crazy aloe people"(CAP). Thus, our project was named the "CAP Project" and it was the beginning of the new Creation of Aloe Pharmaceuticals (CAP).

The CAP project was commenced in June after the visit, and since its beginning, we organized a systematic research team to pursue aloe research. Dr. Seung Ki Lee, a professor at Seoul National University, and twelve other professors from various universities in Korea joined the team. An additional twenty-six professors from prominent universities and their staff have participated in the project.

Many active ingredients from natural aloe have been isolated and analyzed. These conclude that aloe is medically effective in the following cases: promoting liver cell growth, angiogenesis, growth of epithelial cells, anti-allergic effect, skin whitening effect, and protective effect on nephrotoxicity. These results became the groundwork for the development of functional health foods and cosmetics.

Especially, the "development of tolyl-based skin whitening agent" project was proven scientifically reliant, as it was adopted as one of the G7 projects by the Ministry of Science and Technology in Korea. In addition, the "Raman whitening product," which was produced on the basis of the project, was awarded the "IR52 Jang Young-Shil" prize by the Korea Industrial Technology Association and Maeil Economy Newspaper contest in 1998. It was the first case that a cosmetics corporation received an award in the industry. In addition, a study on utilizing an anti-allergic substance, Alprogen, was also adopted as one of the G7 projects and continued for five years. "Development of Prokidin as a protective agent against Cispatin induced nephrotoxicity" was also selected as one of core projects by the Ministry of Health and Welfare. All of these R&D efforts resulted in seven domestic patents and several international pending patents.

After the completion of the first stage of the CAP project, Stage 2 was launched in 2002. The second project focused on reviewing less sufficient tasks of the first project and was more in-depth, aiming to discover new functions of aloe using more advanced methods of verifying aloe's efficacy.

It can be said that the remarkable performance during the past thirteen-year study on aloe is unimaginable, compared to where we were when we initially launched the project. Our systematically designed project has been pursued by a top-down approach and now evaluated to operate in a systematic and efficient way. I think it is a model of a successful R&D project.

I sincerely would like to give special thanks to Professor Seung Ki Lee and other participating researchers who had worked day and night for this project. I would also like to extend my appreciation to the officers and employees of the Namyang Aloe Company that provided much-needed assistance in times of difficulty.

**Park, Young In, Ph.D.**
Editor in Chief
Chairman, The CAP Research and Planning Committee

**Young In Park, Ph.D.**
Korea University, Seoul, Korea

**Seung Ki Lee**
Seoul National University, Seoul, Korea

# Contents

# Contributors

**Chung, Myung Hee, Ph.D.**
College of Medicine, Seoul National University
28 Yeonkun-Dong, Jongno-Ku, Seoul 110-460, Korea

**Choung, Se Young, Ph.D.**
College of Pharmacy, Kyung Hee University
1 Hoegi-dong, Dongdaemun-gu, Seoul 130-701, Korea

**Jo, Tae Hyung**
Unigen, INC.
200-1, Songjung-Ri, Byeongcheon-Myeon, Cheonan-Si,
Chungnam 330-863, Korea

**Kim, Kyeong Ho, Ph.D.**
College of Pharmacy, Kangwon National University
Chuncheon 200-701, Korea

**Kim, Kyu Won, Ph.D.**
College of Pharmacy, Seoul National University
599 Kwanak-ro, Kwanak-gu, Seoul 151-742, Korea

**Kim, Young Shik, Ph.D.**
Natural Products Research Institute
College of Pharmacy, Seoul National University
28 Yeonkun-Dong, Jongno-Ku, Seoul 110-460, Korea

**Lee, Chong Kil, Ph.D.**
College of Pharmacy, Chungbuk National University
48 Gaeshin-dong, Heungduk-gu, Cheongju, Chungbuk 361-763, Korea

**Lee, Seung Ki, Ph.D.**
College of Pharmacy, Seoul National University
599 Kwanak-ro,Kwanak-gu, Seoul 151-742, Korea

**Park, Jeong Hill, Ph.D.**
College of Pharmacy, Seoul National University
599 Kwanak-ro, Kwanak-gu, Seoul 151-742, Korea

**Park, Young In, Ph.D.**
School of Life Sciences and Biotechnology, Korea University
Anam-dong, Sungbuk-gu, Seoul 136-701, Korea

**Son, Byeng Wha, Ph.D.**
Department of Chemistry, Pukyong National University
599-1 Daeyeon 3-d ong, Nam-gu, Busan 608-737, Korea

**Sung, Chung Ki, Ph.D.**
College of Pharmacy, Chonnam National University
300 Yongbong-dong,Buk-gu, Gwangju 500-757, Korea

**Ro, Jai Youl, Ph.D.**
Department of Pharmacology,
Sungkyunkwan University School of Medicine
Cheoncheon-Dong, Jangan-gu, Suwon, Gyeonggi-do 440-746, Korea

# 1. Overview of Aloe study

**Lee, Seung Ki, Ph.D.**

College of Pharmacy, Seoul National University

## 1.1. Research background

### 1.1.1. Historic background of Aloe as folk medicine

Aloe belongs to Liliaceae, the family of perennial tropical plants of African origin. More than 360 species are known worldwide. Species of aloe which have been used as folk medicine include: Curacao Aloe (Aloe barbadensis or Aloe vera), Cape Aloe (Aloe ferox), and Socotra Aloe (Aloe perryi). Records of the use of Aloe vera as folk medicine date to antiquity with an early account from around 1500 B.C. being described in Sumerian Clay Tablets and the papyrus Ebers, stored in Leipzig University. However, the earliest known record of the folk use of aloe is an engraving in an Egyptian temple, which mentions aloe as the sanctuary plant of immortality that is dated to 4000 B.C.

Ancient records in Egypt, Rome, Greece, Arabia, India and China also demonstrate the use of aloe both as a medicine and cosmetic compound. According to these records, Aristotle persuaded the king Alexander to conquer Socotra Island, located off the east coast of Africa, in order to acquire aloe for use in healing the wounds of soldiers. It is also written that Cleopatra used aloe gel to maintain her beauty. This demonstrates that Aloe has been used as medicine and a cosmetic compound from the earliest recorded history.

The first somewhat detailed description of the pharmacological effects of aloe is recorded in "The Greek Herbal", written by Dioscorides in the first century AD. According to such literature, Aloe has multiple pharmacological effects on healing wounds, burns and frostbite, constipation, insomnia, stomach disease, pain, hemorrhoids, itching, headache, hair loss, gum disease, kidney disease, blisters, sunburn and more.

### 1.1.2. Scientific approach of Aloe research and the current status

Once a multipurpose folk medicine, the use of aloe has been dramatically reduced with the rise of the pharmaceutical industry in Western Europe. Due to the fact that the tropical plant could not be cultivated in Western Europe, it joined the ranks of many such plants which were classified as "folk medicine" or "botanicals".

The decline in the popularity of aloe can be traced by counting published research articles related to it. The majority of recently published scientific studies of aloe are concerned with the classification of the plant itself or its components.

Thus, whereas much study has been carried out on the components of the plant, little has investigated which of its components exert a pharmacological effect, suggesting the lack of documentation on any new drug development using aloe.

A heightened re-evaluation of and renewed interest in the efficacy of aloe has arisen recently among many medical researchers involved with ameliorating the effects of x-ray radiation from nuclear weapons on the skin. Dermatological conditions caused by exposure to radiological fallout showed that skin healed quite poorly in comparison with healing from other ordinary skin diseases, and required a much longer period of time. Significant skin ulceration was also noted. Thus, under the auspices of national funding, many scientists have conducted research into these skin problems and one study showed that aloe gel was the most efficient treatment method among many studied therapies in healing skin ulcers induced by nuclear x-ray radiation.

Soon after this report, many scientists became more interested in the efficacy of aloe. C.E. Collins, a physician in Maryland, USA, and his son, Reston Collins, published the first research report on "Alvagel", an ointment made from the leaf of *Aloe vera* which demonstrated excellent healing effects on x-ray induced skin ulcers. Later, J.E. Crewe also reported that *Aloe vera* has healing efficacy for chronic skin ulcer, eczema, burn, sunburn, and allergy. He further reported in consecutive volumes of the Minnesota Journal of Medicine in 1937 and 1938 that the ointment did not leave any scars on its treatment site.

Subsequently F.B. Mandeville reported in 1939 that *Aloe vera* is effective in treating mucous membrane ulcers caused by radioactivity. In addition, Lushbaugh and Hale noted that *Aloe vera* cured skin ulcers that were caused by nuclear B-ray beams of 28,000 rep. over 2-month animal studies. This finding was later reported and acknowledged as correct by the United States FDA (Food and Drug Administration).

In spite of all of these successful research results, aloe did not become a target of new drug development, presumably because of the inability to mass produce the plant and the need to transport fresh material. At that time, fresh aloe was only harvested in tropical regions and transported by ship. This raised the question of how to preserve fresh plants in large quantities. Systematic research of aloe then lapsed.

### 1.1.3. CAP-Project: Creation of Aloe Pharmaceuticals for new drug development and establishment of a research team

As stated above, aloe has a 6000-year history of use as a folk medicine and cosmetic, and its efficacy has been proven by modern scientific methods of clinical research. Although the US FDA has approved its efficacy, aloe still remains merely a botanical agent which as yet remains undeveloped into a medicine, and which has only found application as a food aid, functional food, or raw material for cosmetic purposes.

Recently, the sources which have limited aloe research, namely the problems with mass production and maintenance of fresh material, have been overcome through the ongoing efforts and research conducted by Namyang Aloe Corporation, a Korean company, and Aloecorp, a US based company established by Namyang Aloe.

Aloecorp was founded in 1988 to produce, process, and market aloe. The company owns 980 acres of aloe plantations in the US (Texas) and Mexico, and has established automated production methods of sourcing and processing the completed food GMP manufacturing facilities in line with FDA guidelines. In addition, it has started to provide the fresh raw material for domestic and international consumption. The company is equipped with both the knowledge and facilities to separate and purify the clinically effective components of aloe on an industrial scale following considerable research effort.

As a result, Namyang Aloe Corporation has become was ideally positioned to define the various pharmacological effects of aloe, an ancient folk medicine, and proceeded to advance  in June, 1993 through the creation of a research team (CAP-creation of Aloe Pharmaceuticals) to develop new drugs from the plant. The first phase CAP project lasted until 2001.

## 1.2. Research strategy of CAP

The following conditions were considered prior to attempting the development of new drugs from aloe. First of all, aloe has been continuously used as a folk medicine for 6000 years, allowing the formation of an established reputation of clinical purposes. Secondly, as described earlier, the pharmacological efficacy of aloe as a folk medicine has already been partially proven by modern research and clinical application. Thirdly, the limitations of insufficient fresh plant material for systematic aloe research have now been overcome by modern methods of large-scale cultivation and preservation. Finally, the previous research data of aloe was based on the efficacy of the whole extract of aloe or aloe gel, as opposed to that at the level of the purified single component. It was this shortcoming which provided the focal point of our future research.

Based on this background in drug development from the plant, our research team established its research strategy. First, the pharmacological efficacy of aloe extract, the ancient folk medicine, needed to be verified again with contemporary analytical methods. Secondly, a systematic approach had to be developed to separate and identify the structures of the potentially effective components in the aloe extract in order to determine the pharmacologically therapeutic components. Next, studies were required to determine the toxicity of such components, and if indicated, preclinical and clinical studies of the purified, effective components also needed to be undertaken in order to satisfy the requirements of a new drug. The conduct of such systematic and efficient research aimed at the necessary new drug development led to the establishment of the CAP-research team of distinguished scientists from several Korean universities and institutions nationwide.

### 1.2.1. Organization of the CAP research team

To ensure the efficient performance of such pharmacological research, it was necessary to establish a coordinated research team among experts from various universities and institutions. This group was established in a way to minimize any disharmony that might hamper the research. Given the diverse nature of the group, this was a necessity to ensure success as a whole. Two main teams were constructed. The component separation and analysis team separated different components of aloe and controlled the quality of the Aloe products while the efficacy investigation team studied the function of separated components identified by the aforementioned team. This organic system allowed our research team a broad and systematic study of component separation and confirmation of the efficacy of the separated components by established analysis methods in an efficient manner.

In the first year, the principal task was an extensive review of all research data concerning aloe and related patent records made worldwide by the newly organized committee of experts in related fields. The fruits of the effort were collected in an electronic library and made available as references for every participating researcher. Based on the analysis of the literature, the research context and detailed directions of research were determined. In addition, in order to encourage teamwork among the researchers, the author of an initial study was included as a co-author in the final paper or as an inventor if a patent was made of the separated component of aloe, when found by the component separation and analysis team to match with proven efficacy that carried out by the efficacy investigation team.

Several times per year, data presentations were also encouraged to share research progress among the nationwide group of researchers. Such meetings included exchanging raw data, analyzing results, and finding new directions for future research. The accomplishments thus far have been made possible by the sincere effort of all participating researchers even while busy with their academic duties such as giving lectures and conducting their own research activities.

The component separation and analysis team comprised Professors Sung, Chung Ki (team leader from Cheonnam National University), Park, Young In, Park, Jeong Hill, Kim, Young Shik, Kim, Kyeong Ho, Son, Byung Wha, and Choi, Jae Sue, while the efficacy investigation team comprised Professors Lee, Seung Ki, Chung, Myung Hee, Park, Young In, Kim, Kyu Won, Lee, Chong-Kil, Choung, Se Young, and Ro, Jae Youl, all of whom are established authorities and dignitaries in their own universities and research institutions. This group successfully accomplished the objectives of the first phase of the CAP project.

### 1.2.2. The second phase research of the CAP project

The second phase CAP project was launched in 2002 under the leadership of Park, Young In (Korea University). In contrast to the first phase CAP project, the second phase CAP project concentrated on specific topics that had been selected based on the first phase CAP project. The research topics included the isolation and identification of the molecular entity for the wound-healing activity (Park, Young In), anti-allergic activity (Park, Young In) and anti-Alzheimer's disease activity (Kim, Young Ho), and the identification of optimal molecular size of aloe polysaccharides exhibiting maximum immuno-modulatory activity *in vivo* as well as *in vitro* (Lee, Chong-Kil). The second phase CAP project will continue till the end of 2005.

# 2. The history of Aloe

**Sung, Chung Ki, Ph.D.**

College of Pharmacy, Chonnam National University

## 2.1. History

The name aloe is from the Greek alsos and refers to the bitter juice from the leaves of these plants. It is probably derived from the earlier Arabic word alloeh or the Hebrew word allal, both meaning bitter.

Aloe, the drug, one of the oldest crude drugs, has been used as five types of extract from aloe plants: fresh whole leaf, fresh gel (pulp), juice (sap), juice extract and dried gel. Among them, first three were restricted in their use for the reason of preservation of the plant while the last two have been available in the market. Juice extract was used as a crude drug in West, but in the Orient, having been introduced as a Western crude drug (in contrast to Oriental crude drugs originated from the Orient, mainly China), it has been produced by evaporating water from the juice by boiling, to produce a black, hard and shining mass material, whereas dried gel, as a healthy functioning food and drug, is normally a yellow to brown powder made up usually manufactured by freeze drying. The former was believed to be produced historically due to for the reason of the difficulties in preserving or carrying an aloe before use and preventing deterioration. However, almost all of the aloe used recently is the latter.

From the ecological nature of the aloe plant, aloe originated from the Africa and the history of its use dates back almost 6000 years. In 4000 B. C. the aloe plant was engraved in intaglio on an Egyptian temple fresco. A Sumerian clay tablet from 2200 B. C. was the first document to include aloe among the plants of great healing power. The first detailed description of aloe's medicinal value is probably in the Ebers Papyrus written around 1550 B. C. in Egypt. This document gives twelve formulas for mixing aloe with other agents to treat both internal and external human disorders. One of the uses was as a laxative as well as Senna and Oleum Ricini. Other ancient records describe Hippocrates applying aloe clinically and Cleopatra using an aloe gel as cosmetics. Even in the Bible, aloe is described in several sections. Actually aloe has been used as a drug in north Africa from 20 ~ 30 centuries B. C. Alexander the Great conquered the island of Socotra to get sufficient amounts of aloe to use it as a wound healing agent for this soldiers. In 50 B. C. aloe was introduced into western medicine mainly as a laxative. The first detailed description about the pharmacological effects of aloe was in the Greek Herbal (written by Dioscorides, 41 ~ 68 A. D. It states that the sap, not the gel, was collected and boiled down into a thick black mass for storage and transportation. As for its healing effects, it induces sleep, cleanses the stomach, treats for boils and ulcerated genitals, heals foreskin, and is good for dry itchy skin irritation, hemorrhoids, bruises, stops hair loss, and mouth pain, and stops bleeding of wounds, heals tonsillitis and diseases of the mouth and eyes. Until 1200 aloe plant and dried sap widely accepted healing agents throughout Europe. In 1300 ~ 1500, processed aloe (dried sap) was introduced to English medicine and used as a purgative and as a treatment for external wounds and diseases. Until 1700, aloe and aloe plant were spread throughout the Caribbean islands, Central and South America. Especially in the island of Barbados and Curacao in the Caribbean, aloe plant was developed as a commercial crop. Sap was extracted from the plant and boiled down into a black mass for export, mainly to Europe. In 1720, Carl Von Linne named aloe scientifically as *Aloe vera* L. ("true aloe"). Aloe was officially listed as a purgative and a skin protectant by the United States Pharmacopoeia (U. S. P.) in 1820.

In the Orient, aloe is referred as "Lu-hui" in Chinese, meaning "black deposit", referring to the plant resin condensed black, being wrong in the origin of the word. Aloe was introduced to Asia in 300 ~ 400 B. C. in the course of the Persian Conquest by Alexander the Great to western Asia, via the silk road to China. The first record of aloe in the Chinese literature was in the official Materia Medica of the Song Dynasty (700 ~ 800 A. D.). In that book, the whole leaf of aloe, ground and boiled down to black extract, was cited as the main effective item in the treatment of sinus, fever, skin disease and convulsion in children. However, even before the Song Dynasty, in the Dang Dynasty, the same type of aloe (juice or whole leaf extract) was used as a healing agent both internally and externally. For the aloe plant, *Aloe vera* was first introduced for growth in 1817. In Korea, aloe was referred to as "Nohwe", the same word as the Chinese "Lu-hui", differing only in pronunciation. The introduction of aloe into Korea was uncertain, but considering the history of exchange between China, it must date back to the Dang or Song Dynasties from China. The first record about aloe in a Korean herbal book was in 1610 in Dongeubogam by Hur Joon, in which the records in several Chinese herbal books were introduced. After that, several herbal books or prescription books cited aloe for its precaution of use, clinical applications and prescriptions. Aloe (juice extract) was officially recognized as a drug and cited in the first edition of Korean Pharmacopia (K. P.) in 1958, and has continued to be listed to the most recent edition.

In Japan, Aloe was referred to as "Rokai" or "Roeh", the same word as the Chinese "Lu-hui", differing only in pronunciation. Aloe was introduced by missionaries after the opening of a port to foreign trade. Aloe was cited in several herbal books as a plant resin imported by western people for its bitter taste and for its effectiveness in killing insects. Aloe was officially recognized as a drug and cited in the first edition of Japanese Pharmacopia (J. P.) in 1886. Traditionally, the Japanese called aloe "the plant to make a doctor needless" and cultivated some in the garden. The main species cultivated in Japan is *Aloe arborescens* Mill. var. *natalensis* Berg. ("Kitachi" aloe), which is different from the aloe plant used in the West. They have used fresh leaf juice in the treatment of constipation and gastric disorder internally, and of burns, wounds and abrasions externally.

Aloe, historically used as a laxative in a large dose or a stomachic in a small dose in the form of juice (or whole leaf) extract and as a healing agent of burns, skin ailments and wounds in the form of fresh juice for more than 20 centuries, fell into disuse in the West as the seat of civilization moved to the temperate zones where the tropical plant could not survive the freezing winters. The more common use and rediscovery of aloe had to wait until Dr. Collins (U. S. A.) reported the first medical paper on the new effects of aloe on radiation dermatitis or radiation burn. Radiation burns showed skin ulcerations which were nearly incurable were cured successfully even without gross scar tissue. The applications of aloe to atomic radiation burns as well as the addition of folk medicinal uses like stomach ulcer or constipation have focused scientific attention upon aloe again and have lead to the development of the aloe industry. The Aloe plant has come to be grown in warm regions world wide for commercial production of aloe sap and gel. The first commercial aloe farm was established in Florida, U. S. A. in 1912. In 1959 Mr. Stockton successfully stabilized the healing gel and developed Alo-Creme ointment for the treatment of thermal burn. During the last fifty years, research programs on aloe have been undertaken in many parts of the world. In the West, including Russia which is the leader by far, many intensive studies have been carried out to determine the mechanism and components of aloe's effectiveness. As its effectiveness began to spread, numerous companies were formed to produce aloe products. There is still great expectation that soon everyone will fully appreciate this amazing gift of nature.

## 2.2. Aloe, the drug

To say nothing of the West and the Orient, the drug form of aloe was different from the form of aloe used as functioning health foods or drugs recently. The juice that flowed out from the leaves stacked like books was evaporated by boiling, to form a dark brown to black brown mass, hard and shining material, whereas the recent product was obtained by cutting off the outer layer and freeze drying the inner gel. Aloe products differ in their shape, property and name according to the place of production. The main Aloe products are as follows.

1. Cape Aloe: *Aloe ferox* Mill. is the main plant source. Besides this species, the crossbreds with *A. africana* Mill. or *A. spicata* Baker are also used. The cutinized leaves are cut from the bottom of the leaf and over 200 leaves are stacked like a tent. The exuded leaf juices are collected on sheepskin for 6 hours. The collected leaf juice is heated for 4 hours to be concentrated. This concentrated leaf juice is collected in tin bottles and left to be solidified. This drug is an irregular, massive, solid black brown to green brown material. The crushed surface is shiny like glass and a slice is semitransparent like amber and yellow to yellow-brown in color. Because of its semi-transparency it is called Aloe Lucida.

2. Socotrine Aloe: Made of *Aloe perryi* Baker. A semitransparent, yellow-brown, resinous solid. in which tiny solid crystals of aloin can be seen.

3. Curaçao Aloe: *Aloe vera* L. (= *A. barbadensis* Mill.) and its variety *A. vera* L. var. *chinensis* (Haw.) Berger are the main sources. Collected leaves are stacked

in a V shape and the expressedleaf juice is collected and heated to evaporate the water. The temperature of heating is regulated at a rather low level compared to that of Cape Aloe and the product is not transparent. It becomes an opaque, red-brown to dark brown solid and minute solid crystals of aloin are formed. This aloe, as well as Socotrine Aloe, is called Aloe Hepatica.

4. Natal Aloe: Obtained from *A. bainesii* Th. Dyer mainly, as well as from *A. succotrina* Lam., *A. ferox* Mill. or *A. marlothii* Berger.

5. Bombay Aloe: From *A. perryi* Baker or uncertain. Probably produced in eastern Africa and exported from Bombay via Zanzibar.

## 2.3. Aloe, the plant

Aloe, the plant, belonging to the Lily family (Liliaceae), is native to the Mediterranean region of southern Europe and north Africa. Aloe genus contains over three hundred different species including inter-species crossbreeds which grow in the warm regions of Africa, Europe, Asia and the Americas. They are classified into two groups according to their use: for production of extract as crude drugs and for production of gel as health functioning foods. Among them, only a few species have been used commercially, others are merely of decorative value.

The species for the production of crude drugs are *Aloe vera* L. (= *A. barbadensis* Mill.), *A. ferox* Mill., *A. perryi* Baker, etc., and species for use in health functioning foods are *Aloe vera* L., *A. arborescens* Mill., *A. saponaria,* etc. Only two species are grown today commercially, with *Aloe barbadensis* Miller and *Aloe arborescens* being the most popular.

1. *Aloe vera* L. (= *A. barbadensis* Mill. = *A. vulgaris* Lam., Curaçao aloe, the true aloe): *Aloe vera*, meaning the true aloe, was spread throughout the Mediterranean region by man, so it is difficult to discern where it originated. Some have thought it native to the Canary Islands. Its closest relatives, however, occur in Arabia, and this is its most probable area of origin. It can be found there today, on the lower slopes of the coastal mountains. This species is a large, very short stem rosette perennial succulent. The fleshy, sword-shaped leaves are gray-green and grow to 80 cm long. Younger leaves have pale spots. The plant will slowly offset to form a clump. Only large plants flower. The unbranched flower spike carries yellow, tubular flowers. *Aloe vera* var. *chinense*, a hybrid, has smaller rosettes that offset vigorously and flowers rarely. This plant, coming from lowland, subtropical areas, is not very frost-hardy. It grows well along the coast where frost is rare. It prefers good sun, with some shade at midday. Plant it in a well-drained soil emended with organic matter and sand or expanded shale. Watering can be infrequent once the plant is established.

2. *Aloe arborescens* Mill. (= *Aloe mutabilis*, krantz aloe): The Latin word *arborescens* means tree-forming or tree-like, and is a bit misleading in that this

aloe is not really tree-like, but the name was originally applied to this species in reference to the stem-forming habit. The common name krantz aloe refers to its habitat, a krantz being a rocky ridge or cliff. The species formerly known as *Aloe mutabilis* is now regarded as a synonym of *Aloe arborescens*. It develops into a multi-headed shrub 2 ~ 3 m high with striking grey green leaves arranged in attractive rosettes. The leaf margins are armed with conspicuous pale teeth. The large, colorful flower spikes are borne in profusion during the cold winter months, brightening up a drab winter garden. Deep orange is the most common color, but there are also pure yellow forms, and an unusual bi-colored form of deep orange (almost red) and yellow. The inflorescence is usually unbranched, with two to several arising from a single rosette. As with all the aloes, the flowers produce nectar and are attractive to many kinds of bird, in particular the small and colorful sunbirds, which flit from flower to flower in search of nectar. The flowers also attract bees.

This species is distributed mainly over the eastern, summer rainfall areas of the country. It has the third widest distribution of any aloe, occurring from the Cape Peninsula along the eastern coast, through KwaZulu-Natal, Mpumalanga and Limpopo province and further north into Mozambique, Zimbabwe and Malawi. It is one of the few aloes that can be found growing from sea level right up to the tops of mountains. Although its common name refers to the habitat, it is usually found in mountainous areas where it favors exposed ridges and rocky outcrops. It is also found in dense bush. It is possibly the most widely cultivated aloe in the world and can be seen grown in gardens in many cities around the world.

3. *Aloe saponaria* (Ait.) Haw. (African aloe, soap aloe): *Aloe saponaria* is native to arid regions in eastern South Africa, Botswana and Zimbabwe. It grows in a stemless rosette, and produces little offset rosettes around its margin. The main rosette gets up to about 45 cm tall and just as wide. The lance-shaped leaves are thick and succulent, pale green with white speckles, and 25.4-30.5 cm long. The leaf margins are armed with sharp, dark brown teeth. Throughout much of the summer, it sends up a purplish-branched stalk about 0.6 m tall, bearing showy tubular yellow, orange or red flowers. It is cultivated mainly for decoration.

*Aloe barbadensis Mill*    *Aloe arborescens Mill*    *Aloe saponaria (Ait.) Haw*
(See Plate 1.)

4. *Aloe arborescens* Mill. var. *natalensis* Berg.: This species is used mainly in Japan, as a folklore for the treatment of gastrointestinal ailment, burns, bites, athlete's foot and so on.

5. *Aloe brevifolia* Mill. (Blue aloe): *Aloe brevifolia* grows on small hills and slopes in clay in stony ground. This aloe is a coastal species and is found from Bredarsdorp to the Riversdale area. Occurs in a winter and summer rainfall area (375 mm). A neat, compact aloe of height 40 cm - 50 cm that grows in groups of 10 rosettes of 100 mm in diameter, these are formed by offsets at the base. The leaves are glaucus-green, often tinged pink, with white teeth on the margin. The sap is clear. The inflorescence is 400 mm long with a conical raceme. The racemes are orange-scarlet. Flowers in October and November.

6. *Aloe ferox* Mill. (= *A. supralaevis* Haw., bitter aloe, red aloe): This bitter aloe is most famous for its medicinal qualities. In parts of South Africa, the bitter yellow juice found just below the skin has been harvested as a renewable resource for two hundred years. The hard, black, resinous product is known as Cape aloes or aloe lump and is used mainly for its laxative properties but is also taken for arthritis. "Schwedenbitters" which is found in many pharmacies contains bitter aloe. The gel-like flesh from the inside of the leaves is used in cosmetic products and is reported to have wound healing properties. Interestingly, Aloe ferox, along with Aloe broomii, is depicted in a rock painting which was painted over 250 years ago.
Thus bitter aloe will reach 2-3 meters in height with the leaves arranged in a rosette. The old leaves remain after they have dried, forming a "petticoat" on the stem. The leaves are a dull green, sometimes with a slightly blue look to them. They may also have a reddish tinge. The "A. candelabrum form" has an elegant shape with the leaf tips curving slightly downwards. The spines along the leaf edge are reddish in color. Spines may also be present on upper and lower surfaces of the leaves as well. Young plants tend to be very spiny. The flowers are carried in a large, candelabra-like flower head. There are usually between five and eight branches, each carrying a spike-like head of many flowers. Flower color varies from yellowy-orange to bright red. "A. candelabrum" has six to twelve branches and the flowers have their inner petals tipped with white.

7. *Aloe marlothii* Berger: *Aloe marlothii* is widely distributed throughout the warmer parts of southern Africa. Plants grow on rocky hillsides and on open plains. In its habitat, rain averages 35 to 100 cm per year, falling in summer. Plants are single-stemmed and can reach 6m in height, though 2 to 3m is more common. The stem is densely covered with old, dry leaves. The fleshy leaves can be more than 1m in length. Both upper and lower leaf surfaces are covered with stout, reddish-brown spines. The inflorescence is a many branched panicle up to 60cm tall with 20 to 30 racemes. The racemes are never upright but instead carried horizontally. Flowers range in color from red to gold with various shades of orange being most common.

8. *Aloe bainesii* Th. Dyer: Main source of Natal Aloe.

9. *Aloe africana* Mill.: A crossbreed with *A. vera*. Main source of Cape Aloe.

10. *Aloe descoingsii* G. Reyn.:   Short stem (4 ~ 5 cm). Small leaves are curled up into inside. White protuberance on the leaves. Red flower in winter.

11. *Aloe perryi* Baker : The true Socotrine aloe plant was first described from specimens sent to Kew Gardens by W. Perry, and was subsequently found by Balfour, of Edinburgh, growing abundantly upon the island of Socotra, especially in the limestone tracts, from the sea level to an altitude of 900 m. Along with it, but much less abundant, was a dwarf species with spotted leaves. It resembles in its general habit the Barbados aloe, but differs in its shorter leaves, and especially in its flowers, which are arranged in looser racemes on longer pedicels and have the tube much longer than the segments.

12. *Aloe spicata* Baker: This species of aloe is the one from which the better qualities of the drug aloes are obtained. It is a native of Southern Africa. Stem round, about 1m high; leaves about 60 cm long, wedge-shaped, spreading at the top of the stem; flowers large, white, spiked, bell-shaped.

13. *Aloe variegata* L. (tiger aloe): The tiger aloe is a succulent plant which belongs, like the other very numerous kinds of aloe, to the family of the lily plants (Liliaceae). It is approx. 30 cm high. It consists of fleshy sheets, which are rosettes forming one above the other. The dark-green sheets, which are finely pointedly at the edge, are provided with a light green to white grain, which runs crosswise in each case over the sheet. With good care in culture including room culture the first blooms in reddish colors with long panicle appear after a few years. Their homelands are dry areas in South Africa. The tiger aloe is a plant of easy handle, which prospers problem-free beside a bright window to some extent, if one does not kill it by too much water.

Other species of aloe not mentioned here are listed in Table 1. This large number of species is very important not only in view of resources but also in terms of the varieties of species. It is necessary to investigate species other than the currently used species, botanically, chemically and biologically, in order to develop new sources of aloe having special effectiveness like immune modulating, ulcer treating and whitening and so on.

**Table 1.** Aloe species

| No. | Scientific Name | No. | Scientific Name |
|-----|-----------------|-----|-----------------|
| 1 | *Aloe abyssinica* Lam. | 2 | *A. aculeata* Pole Evans |
| 3 | *A. acutissima* H. Perrier | 4 | *A. adigratana* G. Reyn. |
| 5 | *A. albiflora* Guillaumin | 6 | *A. alooides* (Bolus) Van Druten |
| 7 | *A. ammophila* G. Reyn. | 8 | *A. amudatensis* G. Reyn. |
| 9 | *A. andogensis* Baker | 10 | *A. angelica* Pole Evans |
| 11 | *A. angiensis* | 12 | *A. ankaboensis* nom. nud. (= *A. vinabinensis* nom. nud.) |
| 13 | *A. antandroi* (R.Decary) H.Perrier | 14 | *A. arborescens* var. miller |
| 15 | *A. archeri* Lavranos | 16 | *A. arenicola* G. Reyn. |
| 17 | *A. aristata* Haw. | 18 | *A. audhalica* Lavranos & Hardy |
| 19 | *A. babatiensis* Christian & I Verd. | 20 | *A. baker* Scott Elliot |
| 21 | *A. ballyi* G. Reyn. | 22 | *A. barbertoniae* |
| 23 | *A. bargalensis* Lavranos | 24 | *A. bellatula* G. Reyn. |
| 25 | *A. bergalensis* 26 | 26 | *A. berhana* |
| 27 | *A. branddraaiensis* Groenewald | 28 | *A. broomii* Schonl. |
| 29 | *A. buchananii* Baker | 30 | *A. buchlohii* Rauh |
| 31 | *A. buhrii* Lavranos | 32 | *A. bukobana* |
| 33 | *A. bulbillifera* H. Perrier | 34 | *A. burgersfortensis* G. Reyn. |
| 35 | *A. calidophila* G. Reyn. | 36 | *A. cameronii* Hemsley |
| 37 | *A. camperi* Schweinf. | 38 | *A. candelabrum* Berger |
| 39 | *A. capensis* 40 | 40 | *A. castanea* Schönl. |
| 41 | *A. catengiana* G. Reyn. | 42 | *A. chabaudii* Schönl. |
| 43 | *A. cheranganiensis* Carter & Brandham | 44 | *A. chinensis*(= *A. vera* var. *chinensis*) |
| 45 | *A. chortolirioides* Berger | 46 | *A. christianii* G. Reyn. |
| 47 | *A. chrysostachys* Lavranos & Newton | 48 | *A. ciliaris* Haw. |
| 49 | *A. classenii* G. Reyn. | 50 | *A. commelini* |
| 51 | *A. comosa* Marloth & Berger | 52 | *A. compacta* G. Reyn. |
| 53 | *A. comptonii* Reynolds | 54 | *A. confusa* Engler |
| 55 | *A. cooperii* | 56 | *A. cremnophila* G. Reyn. et Bally |
| 57 | *A. crytopoda* Baker | 58 | *A. davyana* Schönl. |
| 59 | *A. dawei* Berger | 60 | *A. deltoideodonta* Baker |
| 61 | *A. deserti* Berger | 62 | *A. dhufarensis* Lavranos |
| 63 | *A. dichatoma* | 64 | *A. dichotoma* Masson |
| 65 | *A. distans* Haw. | 66 | *A. divaricata* Berger |
| 67 | *A. doei* Lavranos | 68 | *A. dolomitica* Groenewald |
| 69 | *A. dorotheae* Berger | 70 | *A. duckeri* Christian |
| 71 | *A. dumetorum* Mathew & Brandham | 72 | *A. dyeri* Schönl. |

| No. | Scientific Name | No. | Scientific Name |
|-----|-----------------|-----|-----------------|
| 73 | *A. ecklonis* | 74 | *A. elegans* Tod. |
| 75 | *A. elgonica* Bullock | 76 | *A. eminens* G. Reyn. & Bally |
| 77 | *A. erensii* Christian | 78 | *A. erinacea* Hardy |
| 79 | *A. eru* | 80 | *A. excelsa* Berger |
| 81 | *A. falcata* Baker | 82 | *A. ferox* Cape hybrid |
| 83 | *A. ferox* Kenya hybrid | 84 | *A. fibrosa* Lavranos & Newton |
| 85 | *A. flexifolia* Christian | 86 | *A. fleurentinorum* Lavranos & Newton |
| 87 | *A. flexilifolia* | 88 | *A. foetida* nom. nud. |
| 89 | *A. forbesii* Balf. *fil* | 90 | *A. fosteri* Pill. |
| 91 | *A. framesii* L. Bolus | 92 | *A. foetida* |
| 93 | *Aloe fosteri* Pillans | 94 | *A. gariepensis* |
| 95 | *A. gerstneri* | 96 | *A. gilbertii* Ined. |
| 97 | *A. gillilandii* G. Reyn. | 98 | *A. glauca* Mill. |
| 99 | *A. globuligemma* Pole Evans | 100 | *A. gossweileri* G. Reyn. |
| 101 | *A. gracilicaulis* G. Reyn.& Bally | 102 | *A. gracilis* Haw. |
| 103 | *A. graminicola* G. Reyn.104 | 104 | *A. grandidentata* Salm-Dyck. |
| 105 | *A. grata* G. Reyn. | 106 | *A. greatheadii* Schönl. |
| 107 | *A. greenii* Baker | 108 | *A. greenwayi* G. Reyn. |
| 109 | *A. harlana* G. Reyn. | 110 | *A. haworthioides* Baker |
| 111 | *A. heliderana* Lavranos | 112 | *A. hemmingii* G. Reyn. |
| 113 | *A. hendrickxii* G. Reyn. | 114 | *A. hereroensis* Engler |
| 115 | *A. hildebrandtii* Baker | 116 | *A. howmanii* G. Reyn. |
| 117 | *A. humilis* (L.) Miller | 118 | *A. ibitiensis* H. Perrier |
| 119 | *A. immaculata* Pill. | 120 | *A. indica* |
| 121 | *A. inermis* Forsskal | 122 | *A. inyangensis* Christian |
| 123 | *A. isaloensis* | 124 | *A. isoaloensis* H. Perrier |
| 125 | *A. jacksonii* G. Reyn. | 126 | *A. jexblakei* Christian |
| 127 | *A. jucunda* G. Reyn. | 128 | *A. juvenna* Brandham & Carter |
| 129 | *A. karasbergensis* Pillans | 130 | *A. keayi* G. Reyn. |
| 131 | *A. kedongensis* G. Reyn. | 132 | *A. khamiensis* |
| 133 | *A. kilifiensis* Christian | 134 | *A. khamiesensis* Pill. |
| 135 | *A. kirkii* Baker | 136 | *A. komatiensis* |
| 137 | *A. krapohliana* Marloth | 138 | *A. lateritia* Engler |
| 139 | *A.lensayuensis*Lavranos&Newton | 140 | *A. lettyae* G. Reynolds |
| 141 | *A. lineata* (Ait.) Haw. | 142 | *A. lingua* Willd. |
| 143 | *A. littoralis* Baker | 144 | *A. longibracteata* Pole Evans |
| 145 | *A. longistyla* Baker | 146 | *A. lucida* |
| 147 | *A. lutescens* Groenewald | 148 | *A. macleayi* G. Reyn. |
| 149 | *A. macracantha* | 150 | *A. macrocarpa* Tod.. |

| No. | Scientific Name | No. | Scientific Name |
|---|---|---|---|
| 151 | *A. macrosiphon* Baker | 152 | *A. marsabitensis* I.Verd.&Chistian |
| 153 | *A. massawana* G. Reyn. | 154 | *A. mawii* Christian |
| 155 | *A. mayottensis* Berger | 156 | *A. mcloughlinii* Christian |
| 157 | *A. medishiana* G. Reyn. & Bally | 158 | *A. megalacantha* Baker |
| 159 | *A. melanacantha* Berger | 160 | *A .menachensis*(Schweinf.)Blatter |
| 161 | *A. metallica* Engl. & Gilg | 162 | *A. microdonta* Chiov. |
| 163 | *A. microstigma* Salm-Dyck | 164 | *A. millotti* G. Reyn. |
| 165 | *A. milne-redheadii* Christian | 166 | *A. mitriformis* Mill. |
| 167 | *A. moledriana* ined. | 168 | *A. monotropa* I. Verd. |
| 169 | *A. monteiroi* Baker | 170 | *A. morijensis* Carter&Brandham |
| 171 | *A. morogoroensis* Christian | 172 | *A. mubendiensis* Christian |
| 173 | *A. mudenensis* G. Reyn. | 174 | *A. munchii* Christian |
| 175 | *A. mutabilis* Pillans | 176 | *A. mzimbana* Christian |
| 177 | *A. kirkii* Baker | 178 | *A. myeriensis* |
| 179 | *A. myriacantha* (Haw.) Roemer& Schultes | 180 | *A. mzimbana* |
| 181 | *A. ngomeni* | 182 | *A. niebuhriana* Lavranos |
| 183 | *A. nobilis* | 184 | *A. nubigena* Groenew. |
| 185 | *A. nyeriensis* Christian | 186 | *A. nyeriensis* ssp. nyeriensis |
| 187 | *Aloe officinalis* Forsskal | 188 | *A. ortholopha* Christian & Milne Redh. |
| 189 | *A. otallensis* Baker | 190 | *A. pachygaster* Dinter |
| 191 | *A. palmiformis* Baker | 192 | *A. parvibracteata* Schönl. |
| 193 | *A. parvula* Berger | 194 | *A. patersonii* B. Mathew |
| 195 | *A. pearsonii* Schönl. | 196 | *A. peckii* Bally & I. Verd. |
| 197 | *A. peersii* nom. nud. | 198 | *A. peglerae* Schönl. |
| 199 | *A. pendens* Forsskai | 200 | *A. penduliflora* Baker |
| 201 | *A. percrassa* Tod. | 202 | *A. petricola* Pole Evans |
| 203 | *A. petrophylla* Pillans | 204 | *A. pillansii* L. Guthrie |
| 205 | *A. pirottae* Berger | 206 | *A. plicatilis* [L.] Mill. (= *A. linguaeformis* L. f.) |
| 207 | *A. pluridens* Haw. | 208 | *A. polyphylla* Schönl. ex Pill. |
| 209 | *A. pratensis* Baker | 210 | *A. pretoriensis* Pole Evans |
| 211 | *A. pruinosa* G. Reyn. | 212 | *A. pubescens* G. Reyn. |
| 213 | *A. pulcherrima* | 214 | *A. purpurasecns* Haw. |
| 215 | *A. rabaeiensis* Rendle | 216 | *A. ramosissima* Pillans |
| 217 | *A. rauhii* G. Reyn. | 218 | *A. recurvifolia* |
| 219 | *A. reitzii* G. Reynolds | 220 | *A. retrospiciens* G. Reyn & I Verd. |
| 221 | *A. reynoldsii* Letty | 222 | *A. rigens* G. Reyn. & Bally |

| No. | Scientific Name | No. | Scientific Name |
|-----|-----------------|-----|-----------------|
| 223 | *A. rivae* Baker | 224 | *A. rivierei* Lavranos & Newton |
| 225 | *A. rubescens* | 226 | *A. rubrolutea* Schinz |
| 227 | *A. rubroviolacea* Schweinf. | 228 | *A. rupestris* Baker |
| 229 | *A. rupicola* G. Reyn. | 230 | *A. ruspoliana* Baker |
| 231 | *A. schelpei* G. Reyn. | 232 | *A. schimperi* Tod. |
| 233 | *A. schliebnii* Lavranos | 234 | *A. schomeri* Rauh |
| 235 | *A. schweinfurthii* | 236 | *A. scobinifolia* G. Reyn. & Bally |
| 237 | *A. scorpioides* Leach | 238 | *A. secundiflora* Engler |
| 239 | *A. sessiliflora* Pole Evans | 240 | *A. simii* Pole Evans |
| 241 | *A. sinkatana* G. Reyn. | 242 | *A. somaliensis* Walter Watson |
| 243 | *A. species* | 244 | *A. speciosa* Baker |
| 245 | *A. spectabilis* G. Reyn. | 246 | *A. spinasissima* |
| 247 | *A. squarrosa* Baker | 248 | *A. striata* Haw. |
| 249 | *A. striatula* Haw. (= *A. macowanii* Bak.) | 250 | *A. suarezensis* H. Perrier |
| 251 | *A. succotrina* Lam. (= *A. soccotrina* DC.) | 252 | *A. suffulta* G. Reyn. |
| 253 | *A. suprafoliata* Pole Evans | 254 | *A. suzannae* R. Decary |
| 255 | *A. swynnertonii* Rendle | 256 | *A. tenuior* Haw. |
| 257 | *A. thorncroftii* Pole Evans | 258 | *A. thraskii* Baker |
| 259 | *A. tidmarshii* (Schönl.) Muller | 260 | *A. tomentosa* Defl. |
| 261 | *A. tororoana* G. Reyn. | 262 | *A. transvaalensis* Kuntze |
| 263 | *A. trichosantha* Berger | 264 | *A. turkanensis* Christian |
| 265 | *A. tweediae* Christian | 266 | *A. ukambensis* G. Reyn. |
| 267 | *A. vacillans* Forsskal | 268 | *A. vanbalenii* Pillans. |
| 269 | *A. vaombe* | 270 | *A. venusta* G. Reyn. |
| 271 | *A. vera* var. *chinensis* | 272 | *A. vera* var. *littoralis* |
| 273 | *A. verdoorniae* G. Reyn. | 274 | *A. viguieri* H. Perrier |
| 275 | *A. vituensis* Baker | 276 | *A. vogtsii* G. Reyn. |
| 277 | *Aloe volkensii* Engler | 278 | *A. vryheidensis* Groenewald |
| 279 | *A. vulgaris* var. *officinalis* Forsk. | 280 | *A. wickensii* Pole Forsk. |
| 281 | *A. wickensii* var. *lutea* Reynolds | 282 | *A. wilsonii* G. Reyn. |
| 283 | *A. wrefordii* G. Reyn. | 284 | *A. yavellana* G. Reyn. |
| 285 | *A. zebrina* Baker | | |

# 3. Chemical components of Aloe and its analysis

## 3.1. An epitome of chemical components and low molecular compounds

**Park, Jeong Hill, Ph.D. and Kwon, Sung Won, Ph.D.**

College of Pharmacy, Seoul National University

Since it was first recorded in 'Ebers Papyrus' in 1500 BC that aloe was used by humans, Cape Aloe (*A. ferox*), Socotra Aloe (*A. perryi*), and Curacao Aloe *(A. barbadensis* or *A. vera*) have been used as private medicines for wounds or burns and as first-aid medicines for multipurposes in Africa and Europe. Dried aloe juice, which is made from *A. ferox*, *A. africana*, and *A. spicata* and leaves of their mixed breed, is used as a laxative. This short review discusses the chemical components of aloe, especially the phenolic compounds, and the methods for their analysis.

### 3.1.1. Chemical components of aloe

The chemical components of aloe are categorized as follows.

**A. Anthraquinones**

Aloe-emodin (Conner *et al.* 1989; Oudtshoorn *et al.* 1964; Choi *et al.* 1996) is a representative compound found in a variety of aloe genuses. Also reported are chrysophanol, nataloe-emodin (Conner *et al.* 1987), aloesaponarin I (Yagi *et al.* 1974), aloesaponarin II (Yagi *et al.* 1974), laccaic acid D-methylester, deoxy-erythrolaccin, helminthosporin, isoxanthorin, 11-O-rhamnosyl aloe-emodin, and 1,5-dihydroxy-3-hydroxymethylanthraquinones. Crystalline aloe-emodin has a needle shape with orange color. It was reported to have anti-inflammatory activity (Yamamoto *et al.* 1991) and genetic toxicity (Muller *et al.* 1996).

**B. Anthranols**

Mainly found in *A. saponaria*, (R)-aloe chysone (Dagne *et al. et al.*, 1992), aloe-esaponol I (Yagi *et al.* 1997; Dagne *et al.* 1992), aloesaponol II (Yagi *et al.* 1997; Dagne *et al.* 1992), aloesaponol III (Yagi *et al.* 1997; Dagne *et al.* 1992), aloe-saponol IV (Yagi *et al.* 1997), aloesaponol I 6-O-β-D-glucoside, aloesaponol III 6-O-β-D-glucoside, aloesaponol III 8-O-β-D-glucoside, aloesa-ponol III 4-O-β-D-glucoside (Yagi *et al.* 1998), and aloesaponol IV 4-O-β-D-glucoside (Yagi *et al.* 1998) have been reported.

## C. Anthrones

In addition to aloin A & B (Hay *et al.* 1956; Park *et al.* 1997) which are found in various aloe genuses, also reported are 7-hydroxyaloin A & B (Rauwald *et al.* 1982), 10-hydroxyaloin A & B (Okamura *et al.* 1997; Park *et al.* 1997; Dagne *et al.* 1996), 10-C-rhamnosyl aloe-emodin anthrone (Conner *et al.* 1989; Park *et al.* 1995), 5-hydroxyaloin A, (±)-homonataloin, microdontin A & B, 8-O-methyl-7-hydroxyaloin A & B, and 5-hydroxyaloin A 6'-O-acetate (Yagi *et al.* 1998). As a representative compound from aloe, aloin is known to have laxative (Mandal *et al.* 1980; Mapp *et al.* 1970), alcohol metabolism facilitating (Chung *et al.* 1996), and anti-inflammatory (Yamamoto *et al.* 1991) activities. While only B-diastereomer of aloin is produced by an enzyme, A-diastereomer is generated by isomerization of B-diastereomer (Grun *et al.* 1980).

## D. Pyrones

Aloenin (Conner *et al.* 1990; Suga *et al.* 1972; Suga *et al.* 1974(a); Suga *et al.* 1978; Suga *et al.* 1974(b)), aloenin X (Woo *et al.* 1994), aloenin B (Speranza *et al.* 1986; Park *et al.* 1995), aloenin-2''-p-coumaroyl ester, and 10-O-$\beta$-D-gluco-pyranosyl aloenin (Duri *et al.* 2004) have been reported. Especially aloenin, which is a major compound in *A. arborescens*, is known to have anti-inflammatory activity (Yamamoto *et al.* 1991) and inhibition activity of gastric juice secretion (Hirata *et al.* 1978).

## E. Chromones

In addition to aloesin (Yagi *et al.* 1977; Comer *et al.* 1990; Mebe 1987; Conner *et al.* 1990; Haynes *et al.* 1970; Holdsworth *et al.* 1971; Yaun    *et al.* 1991) as a representative compound, there are neoaloesin A (Park *et al.* 1996), 8-C-glucosyl-7-O-methyl-(S)-aloesol, 7-O-methyl-aloesinol (Park MK *et al.* 1997), aloeresin D (Speranza *et al.* 1986; Conner *et al.* 1989), isoaloeresin D (Okamura *et al.* 1996; Meng *et al.* 2004), aloeresin E (Okamura *et al.* 1996), 7-O-methylaloesin, and 7-O-methylaloeresin A (Bisrat *et al.* 2000), among which C-glycosylated chromones are known to be unique compounds of aloe, not having been reported in other plants (Franz *et al.* 1983). In particular, aloesin, a chromone found in various aloe genuses, has become of interest due to its whitening activity (Yagi *et al.* 1987) and growth facilitating activity for liver cells (Lee *et al.* 1997). Neo-aloesin A is a C-glucofuranoside compound isolated from aloe genus for the first time.

## F. Miscellaneous

Chlorinated amides including N-4'-chlorobutylbutyramide Coniine (Blitzke *et al.* 2000) and N,N-dimethylconiine (Blitzke *et al.* 2000), have been reported, along with isoflavone glycosides such as 7-hydroxy,6,3',4'-trimethoxy-5-O-alpha-L-rhamnopyranosyl [1→6] glucopyranoside (Saxena *et al.* 2000) and 5,4'-di-hydroxy 6,7,3',5'-tetramethoxy flavone 5-O-alpha-L-rhamno-pyranosyl-[1→6]-O-beta-D-galacto-pyranoside (Saxena *et al.* 1998). Acemannan, a carbo-hydrate fraction (Lee *et al.* 2001) and aloeride (Pugh *et al.* 2001) and lectin (Tian *et al.* 2001; Akev *et al.* 1999), high molecular weight polysaccharides, have also been reported.

Figs. 1 ~ 11. show the structures of the representative compounds in Aloe.

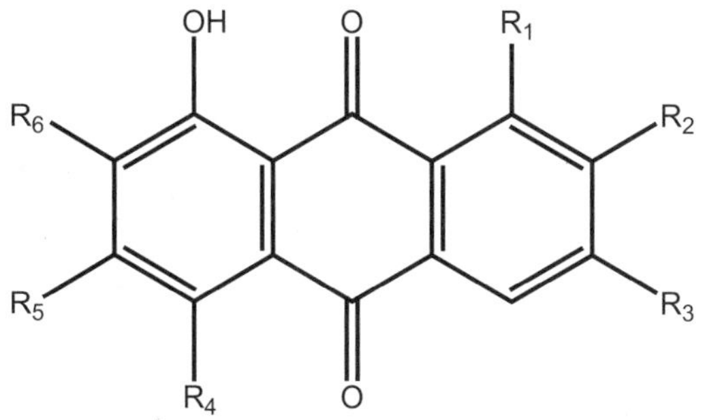

| Compounds | R1 | R2 | R3 | R4 | R5 | R6 | Reference |
|---|---|---|---|---|---|---|---|
| aloe-emodin | OH | H | $CH_2OH$ | H | H | H | 13,16,17 |
| chrysophanol | OH | H | $CH_3$ | H | H | H | 3,4 |
| nataloe-emodin | OH | H | $CH_3$ | H | H | OH | 12 |
| aloesaponarin I | $CH_3$ | COO-$CH_3$ | OH | H | H | H | 2 |
| aloesaponarin II | $CH_3$ | H | OH | H | H | H | 2 |
| laccaic acid D-methylester | OH | COO-$CH_3$ | OH | H | OH | H | 2 |
| Deoxyerythro-laccin | $CH_3$ | H | OH | H | OH | H | 2 |
| helminthosporin | OH | H | $CH_3$ | OH | H | H | 2 |
| isoxanthorin | OH | H | $CH_3$ | OH | H | OC-$H_3$ | 4 |
| 11-O-rhamnosyl -aloe-emodin | OH | H | $CH_2O$-rha | H | H | H | 13 |
| 1,5-dihydroxy-3-hydroxy-methylanthraquinones | OH | H | $CH_2OH$ | OH | H | H | 18 |

**Fig. 1.** Structure of anthraquinones in *Aloe*

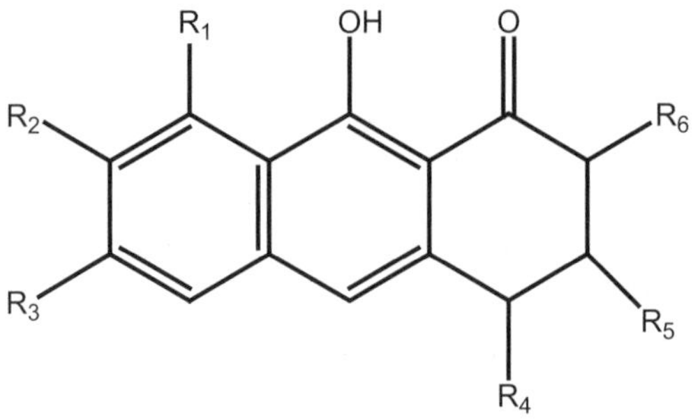

| Compounds | R1 | R2 | R3 | R4 | R5 | R6 | Reference |
|---|---|---|---|---|---|---|---|
| aloesaponol I | CH₃ | COO-CH₃ | OH | H | OH | H | 3,19 |
| aloesaponol II | CH₃ | H | OH | H | OH | H | 3,19 |
| aloesaponol III | OH | H | CH₃ | OH | H | H | 3,19 |
| aloesaponol IV | OH | H | CH₃ | OH | H | OCH₃ | 3 |
| (R)-aloechrysone | OCH₃ | H | H | H | OH, CH₃ | H | 19 |
| aloesaponol I 6-O-β-D-glucoside | CH₃ | COO-CH₃ | O-glc | H | OH | H | 3 |
| aloesaponol III 6-O-β-D-glucoside | CH₃ | H | O-glc | H | OH | H | 3 |
| aloesaponol III 8-O-β-D-glucoside | O-glc | H | CH₃ | OH | H | H | 3 |
| aloesaponol III 4-O-β-D-glucoside | OH | H | CH₃ | O-glc | H | OH | |
| aloesaponol IV 4-O-β-D-glucoside | OH | H | CH₃ | O-glc | H | OCH₃ | |

**Fig. 2.** Structure of anthranols in *Aloe*

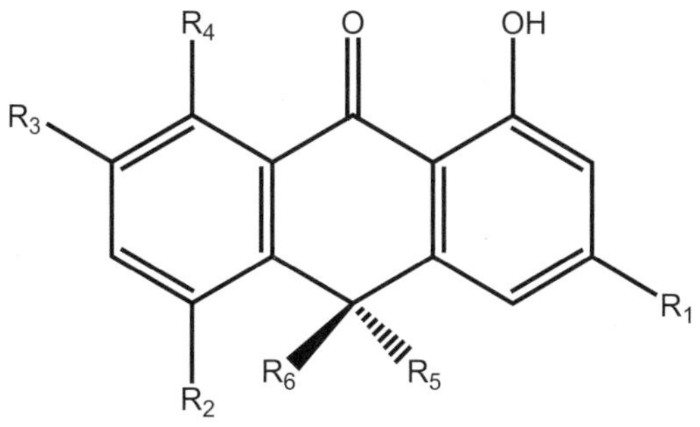

| Compounds | R1 | R2 | R3 | R4 | R5 | R6 | Reference |
|---|---|---|---|---|---|---|---|
| Aloin A | $CH_2OH$ | H | H | OH | H | C-glc | 20,21 |
| Aloin B | $CH_2OH$ | H | H | OH | C-glc | H | 20,21 |
| 5-hydroxyaloin A | $CH_2OH$ | OH | H | OH | C-glc | H | 52 |
| 7-hydroxyaloin A | $CH_2OH$ | H | OH | OH | H | C-glc | 9 |
| 7-hydroxyaloin B | $CH_2OH$ | H | OH | OH | C-glc | H | 9 |
| Aloinoside A | $CH_2O$-α-L-rha | H | H | OH | H | C-glc | 53 |
| Aloinoside B | $CH_2O$-α-L-rha | H | H | OH | C-glc | H | 53 |
| (+)-homonataloin | $CH_3$ | H | OH | $OCH_3$ | H | C-glc | 14,18,54 |
| (-)-homonataloin | $CH_3$ | H | OH | $OCH_3$ | C-glc | H | 14,18,54 |
| Microdontin A | $CH_2OH$ | H | H | OH | H | C-glc-2'-couma-royl | 55 |
| Microdontin B | $CH_2OH$ | H | H | OH | C-glc-2'-couma-royl | H | 55 |
| 10-hydroxyaloin A | $CH_2OH$ | H | H | OH | OH | C-glc | 5,23 |
| 10-hydroxyaloin B | $CH_2OH$ | H | H | OH | C-glc | OH | 22 |
| 8-O-methyl-7-hydroxyaloin A | $CH_2OH$ | H | OH | $OCH_3$ | H | C-glc | |
| 8-O-methyl-7-hydroxyaloin B | $CH_2OH$ | H | OH | $OCH_3$ | C-glc | H | |
| 5-hydroxyaloin A 6'-O-acetate | $CH_2OH$ | OH | H | OH | C-glc-6'-acetyl | H | |

**Fig. 3.** Structure of anthrones in *Aloe*

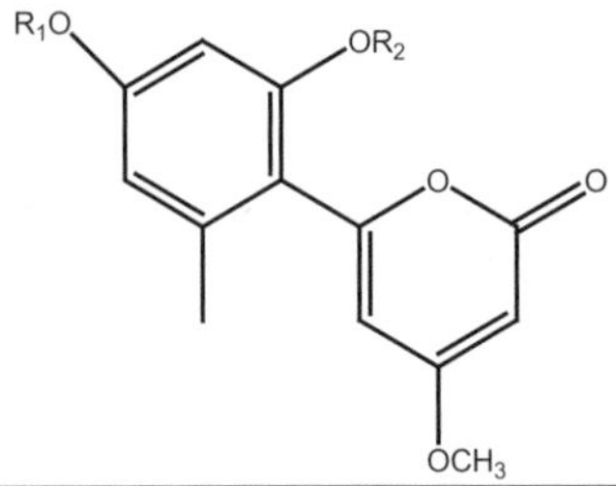

| Compounds | R1 | R2 | Reference |
|---|---|---|---|
| aloenin | H | β-D-glucosyl | 25-29 |
| aloenin B | β-D-glucosyl | β-D-glucosyl-2''-p-coumaroyl | 8,24 |
| aloenin-2''-p-coumaroyl ester | H | β-D-glucosyl-2''-p-coumaroyl | 12 |
| aloenin X | H | 4',6'-O-ethylidene- β-D-glucosyl | 30 |
| 10-O-β-D-glucopyranosyl aloenin | β-D-glucosyl | β-D-glucosyl | |

**Fig. 4.** Structure of pyrones in *Aloe*

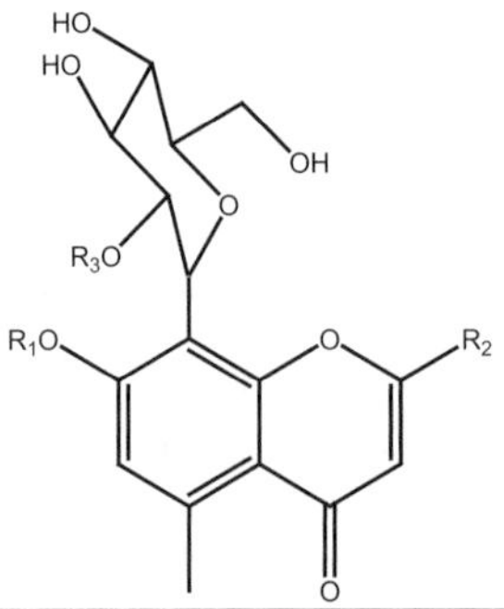

| Compounds | R1 | R2 | R3 | Reference |
|---|---|---|---|---|
| aloesin | H | $CH_2COCH_3$ | H | |
| 8-C-glucosyl-7-O-methyl-(S)-aloesol | $CH_3$ | $CH_2COCH_3$ H OH | H | |
| isoaloeresin D | $CH_3$ | $CH_2COCH_3$ H OH | p-(E)-coumaroyl | |
| aloeresin E | $CH_3$ | $CH_2COCH_3$ H OH | (E)-cinnamoyl | |
| 7-O-methylaloeresin A | $CH_3$ | $CH_2COCH_3$ | p-(E)-coumaroyl | |
| 7-O-methylaloesin | $CH_3$ | $CH_2COCH_3$ | H | |

**Fig. 5.** Structure of chromones in *Aloe*

| Compounds | R1 | R2 | Reference |
|---|---|---|---|
| littoraloside | O-glc | OH | |
| littoraloin | OH | OAc | |
| deacetyllittoraloin | OH | OH | |

**Fig. 6.** Structure of C,O-diglucosylated oxanthrones in *Aloe* 68)

**Fig. 7.** Structure of neoaloesin A

**Fig. 8.** Structure of aloeresin H 64)

**Fig. 9.** Structure of 6'-O-coumaroylaloesin 67)

**Fig. 10.** Structure of 4-methyl-6,8-dihydroxy-7H-benz[d,e]anthracen-7-one 63)

**Fig. 11.** Structure of 3,4-dihydro-6,8-dihydroxy-3-(2'-acetyl-3'-hydroxy-5'-methoxy) methyl-1H-[2]benzopyran-1-one 70)

### 3.1.2. Analysis of components

For the analysis of the chemical components in aloe, various methods such as fluorophotometry (Ishii *et al*. 1984; Vargas *et al*. 2004; Zhao *et al*. 1998), TLC (Wawrznowicz *et al*. 1994; Reynolds *et al*. 1985; Deng *et al*. 1999; Wang *et al*. 2002), size exclusion chromatography (Turner *et al*. 2004), GC (Nakamura *et al*. 1989; Kim *et al*. 1996), GC/MS (ElSohly *et al*. 2004), HPLC (Rauwald *et al*.1993(a); Rauwald *et al*. 1993(b); Suzuki *et al*. 1986, Rauwald *et al*. 1994; Park *et al*. 1998), LC/MS (Rebecca *et al*. 2003; Li *et al*. 2000; Schmidt *et al*. 2001), atomic absorption spectrometry (Xie 2001), counter current chromatography (Yuan *et al*. 2003), capillary electrophoresis (Wang *et al*. 2001), and micellar electro-kinetic chromatography (Kuzuya *et al*. 2001) have been used. However, only the HPLC method has been widely applied to analyze the components in aloe. In the fluorophotometry method, only aloin A and aloin B were detected after Schoutelen's reaction. The TLC method requires a variety of mobile phases and color developing reagents. Size exclusion chromatography lacks high resolution and sensitivity. In the GC method, it is difficult to silylate anthrones and chromones. In addition, they are easily degraded at high temperature. Though other analytical methods such as atomic absorption spectrometry, counter current chromatography, capillary electrophoresis, and micellar electrokinetic chromatography are also available, their application is not practical. On the other hand, HPLC can be directly applied to analyze the components of aloe without further treatment. There are, however, few reports on the analysis of a limited number of phenolic compounds by HPLC, and these only focus on acquiring chemotaxonomic data on anthrones, observing the chromatographic patterns, or quantifying aloin A and aloin B in aloe powder.

Our group has analyzed 13 phenolic compounds, aloesin (Comer *et al*. 1990), 8-C-glucosyl-7-O-methyl-(S)-aloesol (Okamura *et al*. 1996), neoaloesin A (Park *et al*. 1996), 8-O-methyl-7-hydroxyaloin A&B (Okamura *et al*. 1997), 10-hydroxyaloin A (Park *et al*. 1997), isoaloeresin D (Okamura *et al*. 1996), aloin A&B (Comer *et al*. 1990), aloeresin E (Okamura *et al*. 1996), aloe-emodin (Conner *et al*. 1989), aloenin (Hirata *et al*. 1978), and aloenin B (Park *et al*. 1995), using HPLC from *A. barbadensis* and *A. arbo-rescens*, which are widely used as raw materials for health-supporting foods and cosmetics.

### A. Analysis condition

The HPLC system is composed of a Nova-Pak C18 HPLC column (150 mm X 3.9 mm I. D., 4 μm, Waters, USA), Hitachi L-6100 pump (Hitachi, Japan) equipped with Rheodyne 7125 injector (Rheodyne, USA), Hitachi L-4000 UV detector (Hitachi, Japan), and Shimadzu C-R4A Chromatopac integrator (Shimadzu, Japan). Gradient elution using methanol-water gradient, which showed better resolution than acetonitrile-water gradient, was used to separate and quantify the 13 phenolic compounds of various hydrophobicities at the same time. The HPLC gradient condition ranged from 25% methanol-water to 70% methanol-water at a constant mobile phase flow rate of 0.7ml/min. All compounds were clearly separated except the overlapping aloenin and neoaloesin A. Because aloenin is found only in *A. arborescens* while neoaloesin A is in *A. barbadensis*, the overlapping peaks of the two compounds were not a matter of concern. Each peak was identified by comparison of retention time and co-injection of standards. Elution was detected at 293 nm, which is close to the wavelength of maximal absorption for chromones and anthrones. The calibration curves for each compound showed good linearity under the analytical conditions described above.

### B. Determination of phenolic compounds in aloe

*A. barbadensis* showed high content of chromones such as aloesin, 8-C-glucosyl-7-O-methyl-(S)-aloesol, and isoaloeresin D, while *A. arborescens* showed high content of pyrones such as aloenin and aloenin B. Aloins A and B were found at high contents in both species. Aloe-emodin and 10-hydroxyaloin A were also found in both species.

### C. Seasonal change of the contents of phenolic compounds in aloe

In *A. barbadensis*, aloin A & B, 8-O--methyl-7-hydroxyaloin A & B, and 10-hydroxyaloin A showed a large variation by season while chromones such as aloesin and neoaloesin A showed a relatively small change. The content of 8-C-glucosyl-7-O--methyl-(S)-aloesol increased in spring and summer and decreased in fall and winter. In *A. arborescens*, aloenin, aloin A & B, and 10-hydroxyaloin A did not show much seasonal change except in February. The content of Aloenin B was a minimum in March, after which it kept increasing.

### D. Quality control of aloe

Although aloin (barbaloin), which is the representative compound in aloe, is presently used as the marker for quality control (Ishii *et al.* 1984; Zonta *et al.* 1995), it is more reasonable to consider aloesin as a new maker when *A. barbadensis* is used as a raw material, for the following reasons. First, the content of aloesin is consistent across. Second, aloesin is more stable than aloin in acidic and basic conditions and at high temperature (Haynes *et al.* 1970). Third, it is easier to quantify aloesin than aloin (Kim *et al.* 1996). Fourth, aloesin is found only in aloe whereas aloin is reported to exist in other plants (Fairbairn *et al.* 1960; Franz *et al.* 1983). In the case of *A. arborescens*, aloesin could be a good marker for quality control as the content of aloesin is the highest and the seasonal variation of its content is low.

### *3.1.3. Conclusion*

The chemical components from the plants of aloe genuses and their analyses were simply discussed. Between the late 1970s and the mid 1980s, the research on aloe was aimed at developing health foods and cosmetics. Since the 1990s, however, researches have focused mainly on developing new drugs. This trend of research is expected to continue for a while, in accordance with the effort to develop new drugs from natural products. In addition, analyzing the components in aloe using HPLC/mass spectrometry will contribute to the research on aloe.

**References**
1. (1992) The society for pharmacognosy research: modern biology; 15
2. Akev N, Can A (1999) Separation and some properties of Aloe vera L. leaf pulp lectins. Phytother Res 13, 489
3. Beaumont J, Reynolds T, Vaughan J. G. (1984) Homonataloin in Aloe species.
   Planta Med 50, 505
4. Bisrat D, Dagne E, van Wyk BE, Viljoen A (2000) Chromones and anthrones from Aloe marlothii and Aloe rupestris. Phytochemistry 55, 949
5. Bisrat D, Dagne E, van Wyk, BE, Viljoen A (2000) Chromones and anthrones   from Aloe marlothii and Aloe rupestris. Phytochemistry 55, 949
6. Blitzke T, Porzel A, Masaoud M, Schmidt J (2000) A chlorinated amide and  piperidine alkaloids from Aloe sabaea. Phytochemistry 55, 979
7. Choi J, Lee S, Sung C, Jung J (1996) Phytochemical study on Aloe vera. Arch. Pharm Res 19, 163
8. Chung JH, Cheong JC, Lee JY, Roh, HK, Cha, YN (1996) Acceleration of the Alcohol Oxidation Rate in Rats with Aloin, a Quinone Derivative of Aloe. Biochemical Pharmacology 55, 1461
9. Conner JM, Gray AI., Reynolds T, Waterman, PG (1987) Anthraquinone,anthrone and phenylpyrone components of Aloe nyeriensisvar. kedongensis leaf exudate Phytochemistry 26, 2995
10. Conner JM, Gray AI, Reynolds T, Waterman PG (1989) Anthracene and chromone derivatives in the exudate of Aloe rabaiensis Phytochemistry 28, 3551
11. Conner JM, Gray AI, Waterman PG, Reynolds T (1990) Novel anthrone-anthraquinone dimers from Aloe elgonica J Nat Prod 53, 1362
12. Conner JM, Gray AI, Reynolds T, Waterman PG (1990) Anthrone and chromone components of Aloe cremnophila and A jacksonii leaf exudates Phytochemistry 29, 941
13. Dagne E, Casse I, Steglich W (1992) Aloechrysone a dihydroanthracenone from Aloe berhana Phytochemistry 31, 1791
14. Dagne E, Van Wyk B, Stephenson D, Steglich W (1996) Three Oxanthrones from Aloe littoralis Phytochemistry 42, 1683
15. Dagne E, Bisrat D, Codina C, Bastida JA (1998) O-diglucosylated oxanthrone from Aloe littoralis Phytochemistry 48, 903
16. Dagne E, Bisrat D, Van Wyk B, Viljoen A, Hellwig V, Steglich W (1997) Anthrones from Aloe microstigmna Phytochemistry 44, 1271
17. Deng Y, Qin W (1999) Determination of aloe emodin from aloe extract by thin-layer chromatography Shipin Kexue 20, 57
18. Duri L, Morelli CF, Crippa S, Speranza G (2004) 6-Phenylpyrones and 5-methylchromones from Kenya aloe Fitoterapia 75, 520
19. ElSohly MA, Gul W, Murphy TP (2004)Analysis of the anthraquinones aloe-emodin and aloin by gas chromatography/mass spectrometry Int Immunopharmacol 4, 1739
20. Fairbairn JW, Simic S (1960)Vegetable purgatives containing anthracene

derivatives. Part XI. Further work on the aloin-like substance of Rhamnus purshiana DC J Pharm Pharmacol 12, 45T

21. Franz G, Gruen M (1983) Chemistry, Occurrence and Biosynthesis of C-glycosyl compounds in Plants Planta Med 47, 131

22. Farah MH, Andersson R, Samuelsson G (1992) Microdontin A and B: two new aloin derivatives from Aloe microdonta. Planta Med 58, 88

23. Groom QJ, Reynolds T (1987)Barbaloin in Aloe species Planta Med 53, 345

24. Grun M, Franz G (1980) Studies on the biosynthesis of aloin in Aloe arborescens Planta Med 39, 288

25. Gramatica P, Monti D, Speranza G, Manitto P (1982) Aloe revised The structure of aloeresin A Tetrahedron Lett 23, 2423

26. Hay JE, Haynes LJ (1956) The aloins Part I The structure of barbaloin. J Chem Soc 3141

27. Haynes LJ, Holdsworth DK, Russell R (1970) C-glycosyl compounds VI Aloesin a C-glucosylchromone from Aloe sp J Chem Soc (C) 18, 2581

28. Holdsworth DK (1971) Chromones in Aloe species 1 Aloesin C-glucosyl-7-hydroxychromone Planta Med 19, 322

29. Hirata T, Suga T (1978) Structure of aloenin a new biologically-active bitter glucoside from Aloe arborescens var natalensis Bull Chem Soc Jpn 51, 842

30. Horhammer L, Wagner H, Bittner G (1964) Aloinosid B ein neue Glycosid aus Aloe Z Naturforsch 19b, 222

31. Holdsworth DK (1972) Chromones in aloe species II Aloesone Planta Med 22, 54

32. Ishii Y, Tanizawa H, Takino Y (1984) Fluorophotometry of barbaloin in Aloe Chem Pharm Bull 32, 4946

33. Kim KH, Kim HJ, Park JH, Shin YG (1996) Determination of aloesin in aloe preparations by HPLC Yakhak Hoeji 40, 177

34. Kuzuya H, Tamai I, Beppu H, Shimpo K, Chihara T (2001) Determination of aloenin barbaloin and isobarbaloin in Aloe species by micellar electrokinetic chromatography J Chromatogr B 752, 91

35. Lee KY, Park JH, Chung MH, Park YI, Kim KW, Lee YJ, Lee SK (1997) Aloesin up-regulates cyclin E/CDK2 kinase-activity via inducing the protein-levels of cyclin E CDK2 and CDC25A in SK-HEP-1 cells Biochemistry and Molecular Biology International 41, 285

36. Li WK, Chan CL, Lueng HW (2000) Liquid chromatography-atmospheric pressure chemical ionization mass spectrometry as a tool for the characterization of anthraquinone derivatives from Chinese herbal medicine J Pharm Pharmacol 52, 723

37. Lee JK, Lee MK, Yun YP, Kim Y, Kim JS, Kim YS, Kim K, Han SS, Lee CK (2001) Acemannan purified from Aloe vera induces phenotypic and functional maturation of immature dendritic cells Int Immunopharmacol 1, 1275

38. Mapp RK, McCarthy TJ (1970) Assessment of purgative principles in aloes Planta Med 18, 361

39. Muller SO, Eckert I, Lutz WK, Stopper H (1996) Genotoxicity of the Laxative Drug Components Emodin Aloe-emodin and Danthron in Mammalian-Cells-Topoisomerase-II Mediated. Mutation Research-Genetic Toxicology 371, 165

40. Mandal G (l980) Das A Characterization of the polysaccharides of Aloe barbadensis Part I Structure of the D-galactan isolated from Aloe barbadensis Miller Carbohydr Res 86, 247

41. Mebe PP (1987) 2'-p-Methoxycoumaroylaloeresin a C-glucoside from Aloe excelsa Phytochemistry 26, 2646

42. Makino K, Yagi A, Nishioka I (1974) Constituents of Aloe arborescens var natalensis II Structure of two new aloesin esters Chem Pharm Bull 22, 1565

43. Manitto P, Speranza G, De Tommasi N, Ortoleva E, Morelli CF (2003) Aloeresin H a new polyketide constituent of Cape aloe Tetrahedron 59, 401

44. Meng Y, Yan BZ, Wang HM, Hu GF, Liu FY, Song YG, Liu Y (2004) Complete H-1 and C-13 assignments of 8-C-beta-D[2-O-(E)-p-coumaroyl]glucopyranosyl-2-(2-hydroxy)propyl-7-methoxy-5-methylchromone Magn Reson Chem 42, 564

45. Nakamura H, Kan T, Kishimoto K, Ikeda K, Amemiya T, Ito K, Watanabe Y (1989) Gas chromatographic and mass spectrometric determination of aloe components in skin-care cosmetics Eisei Kagaku 35, 2l9

46. Okamura N, Hine N, Tateyama Y, Nakazawa M, Fujioka T, Mihashi K, Yagi A (1997) Three Chromones of Aloe vera Leaves Phytochemistry 45, 1511

47. Okamura N, Hine N, Harada S, Fujioka T, Mihashi K, Yagi A (1996) Three Chromone Components from Aloe vera Leaves Phytochemistry 43, 492

48. Oudtshoorn RV, Van MCB (1964) Chemotaxonomic investigations in Asphodeleae and Aloineae Phytochemistry 3, 383

49. Park MK, Park JH, Shin YG, Choi SM, Choi YS, Kim KH, Lee SK (1997) Chemical Constituents in Aloe barbadensis Arch Pharm Res 20, 88

50. Park MK, Park JH, Kim KH, Shin YG, Myoung KM, Lee JH (1995) Chemical Constituents of Aloe capensis Kor J Pharmacogn 26, 244

51. Park MK, Park JH, Shin YG, Kim WY, Lee JH, Kim KH (1996) Neoaloesin A: A New C-Glucofuranosyl Chromone from Aloe barbadensis Planta Med 62, 363

52. Park MK, Park JH, Shin YG, Choi SM, Choi YS, Kim KH, Lee SK (1997) Chemical Constituent of Aloe capensis Arch Pharm Res 20, 194

53. Park MK, Park JH, Kim NY, Shin YG, Choi YS, Lee JG, Kim KH Lee SK (1998) Analysis of 13 phenolic compounds in aloe species by high performance liquid chromatography Phytochem Analysis 9, 186

54. Pugh N, Ross SA, ElSohly MA, Pasco DS (2001) Characterization of aloeride, a new high-molecular-weight polysaccharide from Aloe vera with potent immunostimulatory activity J Agr Food Chem 49, 1030

55. Rauwald HW, Voetic R (1982) 7-hydroxyaloin: die Leitsubstanz aus Aloe barbadensis in der Ph Eur III Archiv der Pharmazie 315, 477

56. Rauwald HW, Beil A (1993(a)) High-performance liquid chromatographic

separation and determination of diastereomeric anthrone-C-glycosyls in Cape aloes J Chromatogr 639, 359

57. Rauwald HW, Beil A (1993(b)) 5-Hydroxyaloin A in the genus Aloe Thin layer chromatographic screening and high performance liquid chromatographic determination Z Naturforsch., C: Biosci 48, 1

58. Rauwald HW, Sigler A (1994) Simultaneous determination of 18 polyketides typical of Aloe by HPLC and photodiode array detection Phytochemical Analysis 5, 266

59. Reynolds T (1985) Observations on the phytochemistry of the Aloe leaf-exudate compound Botanical Journal of the Linnean Society 90, 179

60. Rebecca W, Kayser O, Hagels H, Zessin KH, Madundo M. Gamba N (2003) The phytochemical profile and identification of main phenolic compounds from the leaf exudate of Aloe secundiflora by high-performance liquid chromatography-mass spectroscopy Phytochem. Analysis 14, 83

61. Speranza G, Dada G, Lunazzi L, Gramatica P, Manitto P (1986) Studies on Aloe Part 3 A C-glucosylated 5-methylchromone from Kenya aloe Phytochemistry 25, 2219

62. Speranza G, Dada G, Lunazzi L, Gramatica P, Manitto P (1986) Studies on Aloe Part 4 Aloenin B a new diglucosylated 6-phenyl-2-pyrone from Kenya Aloe J Nat Prod 49 800

63. Suga T, Hirata T, Odan M (1972) Aloenin a new bitter glucoside from Aloe species Chem Lett 7, 547

64. Suga T, Hirata T, Tori K (1974(a)) Structure of aloenin a bitter glucoside from Aloe species Chem Lett 7, 715

65. Suga T, Hirata T (1978) Biosynthesis of aloenin in Aloe arborescens var natalensis Bull Chem Soc Jpn 51, 872

66. Suga T, Hirata T, Koyama F, Murakami E (1974(b)) Biosynthesis of aloenin in Aloe arborescens var Natalensis Chem Lett 8 873

67. Suzuki Y, Morita T, Haneda M, Ochi K, Shiba M (1986) Determination by high-performance liquid chromatography and identification of barbaloin in aloe Iyakuhin Kenkyu 17, 984

68. Speranza G, Martignoti A, Manitto P (1988) Studies on aloe Part 5 Iso-aleoresin A a minor constituent of Cape aloe J Nat Prod 51, 588

69. Speranza G, Gramatica P, Dada G, Manitto P (1985) Studies on aloe Part 2 Aloeresin C a bitter C,O-diglucoside from Cape aloe Phytochemistry 24, 1571

70. Schmidt J, Blitzke T, Masaoud M (2001) Structural investigations of 5-methylchromone glycosides from Aloe species by liquid chromatography/electrospray tandem mass spectrometry Eur J Mass Spectrom 7, 481

71. Saxena VK (2000) 7-hydroxy,6,3',4'-trimethoxy-5-O-alpha-L-rhamnopyranosyl [1→6] glucopyranoside of Aloe vera Journal of the Institution of Chemists 72, 195

72. Saxena VK, Sharma DN (1998) 5,4'-dihydroxy 6,7,3',5'-tetramethoxy flavone 5-O-alpha-L-rhamno-pyranosyl-[1→6]-O-beta-D-galacto-pyranoside

from Aloe barbadensis (leaves) Journal of the Institution of Chemists 70, 179

73. Turner CE, Williamson DA, Stroud PA, Talley DJ (2004) Evaluation and comparison of commercially available Aloe vera L products using size exclusion chromatography with refractive index and multi-angle laser light scattering detection Int Immunopharmacol 4, 1727

74. Tian B, Wang G, Fang H (2001) Study of purification of lectins from species of Aloe Dalian Ligong Daxue Xuebao 41, 296

75. van Heerden FR, Viljoen AM, van Wyk BE (2000) 6 '-O-coumaroylaloesin from Aloe castanea - a taxonomic marker for Aloe section Anguialoe Phytochemistry 55, 117

76. Vargas F, Rivas C, Medrano M (2004) Interaction of emodin aloe-emodin and rhein with human serum albumin A fluorescence spectroscopic study. Toxicol Mech Method 14, 227

77. Wang HM, Shi W, Xu YK, Wang P, Chen W, Liu Y, Lu MJ, Pa JQ (2003) Isolation and spectral study of 4-methyl-6,8-dihydroxy-7H-benz[de]anthracen-7-one Magn Reson Chem 41, 301

78. Wawrznowicz T, Hajnos MW, Mulak-Banaszek K (1994) Isolation of aloine and aloeemodine from Aloe (Liliaceae) by micropreparative TLC J Planar Chromatogr-Mod TLC 7, 315

79. Wang HM, Shi W, Xu YK, Liu Y, Lu MJ, Pan JQ (2003) Spectral study of a new dihydroisocoumarin Magn Reson Chem 41, 718

80. Wang XP, Ma MX, Shuang SM, Zhang Y, Pan JH (2002) Determination of formation constants for the inclusion complexes between emodin aloe-emodin and cyclodextrins by thin layer chromatography Chinese J Anal Chem 30, 38

81. Wang DX, Yang GL, Song XR, (2001) Determination of pKa values of anthraquinone compounds by capillary electrophoresis Electrophoresis 22, 464

82. Woo WS, Shin KH, Chung HS, Lee JM, Shim, CS. (1994) Isolation of an aloenin-acetal from aloe Kor J Pharnlacog 25, 307

83. Xie LQ (2001) Determination of manganese, iron, zinc, copper, nickel and cobalt in aloe by flame atomic absorption spectrophotometry Chinese J Anal Chem 29, 489

84. Yamamoto M, Masui T, Sugiyama K, Yokota M, Nakagomi K, Nakazawa H (1991) Anti-inflammatory active constituents of Aloe arborescens Miller Agric Biol 5, 1627

85. Yagi A, Makino K, Nishioka I (1974) Constituents of Aloe saponaria. I; Structures of Tetrahydroanthracene Derivatives and the Related Anthraquinones Chem Pharm Bull: 22, 1159

86. Yagi A, Makino K, Nishioka I (1977) Studies on the constituents of Aloe saponaria Haw. II; The structures of Tetrahydroanthracene derivatives aloesaponol III and-IV. Chem Pharm Bull 25, 1764

87. Yagi A, Makino K, Nishioka I (1977) Studies on the constituents of Aloe saponaria Haw III. The structures of phenol glucosides Chem Pharm Bull 25,1771

88. Yagi A, Hine N, Asai M, Nakazawa M, Tateyama Y, Okamura, N., Fujioka T, Mihashi K, Shimomura K (1998) Tetrahydroanthracene glucosides in callus tissue from Aloe barbadensis leaves Phytochemistry 47, 1267

89. Yagi A, Kanbara T, Morinobu N (1987) Inhibition of mushroom-tyrosinase by Aloe extract Planta Med 53, 515

90. Yaun A, Kang S, Tan L, Raun B, Fan Y (1991) Isolation and identification of aloesin from the leaves of Aloe vera L var chinensis (Haw) Berger 31, 251

92. Zhao HC, Feng RQ, Deng XG, Jin LP (1998) Study of the Eu(III)-barbaloin-CTAB system by fluorescence and determination of barbaloin Anal Lett 31, 819

93. Zonta F, Bogoni P, Masotti P, Micali G (1995) High-performance liquid chromatographic profiles of aloe constituents and determination of aloin in beverages with reference to the EEC regulation for flavoring substances J Chromatogr A 718, 99

## 3.2. Proteins in Aloe

**Park, Young In, Ph.D.[1] and Son, Byeng Wha, Ph.D.[2]**

*Aloe vera* contains over 95% water but less than 0.1% protein. Although the amount of total proteins present in aloe is relatively small, its biological activities are meaningful as evidenced by its many clinical applications. The surface layer of aloe leaves (the layer between the gel and the outer epidermis, referred as the 'rind') is where carbohydrates, lipids, proteins and other low molecular weight compounds which contain various physiological activities are produced. One of which found in the rind is carboxypeptidase, a glycoprotein with a molecular weight of either 28 KDa or 56 KDa. It has been proven that this protein is effective in relieving pain caused by inflammation. It has also been reported that the lectin extracted from various types of aloe, including *Aloe vera,* possesses a high level of pharmacological and physiological activities such as mitotic stimulation of lymphocytes, activation of complements by alternative pathway, anti-inflammation, anti-ulcer and anti-tumor. *Aloe arborescence* has been reported to contain 2 types of lectins: aloctin A and B. Aloctin A is composed of subunits α (M.W., 7,5 KDa) and β (M.W., 10,5 KDa) which are linked through a disulfide bond to form a complex of 12 KDa molecular weight, whereas aloctin B has a molecular weight of 24 KDa and is composed of a γ-subunits (M.W., 12 KDa). Other types of protein such as trypsin-like enzymes and protease have been also found in aloe (Fujita *et al.* 1976; Fujita *et al.* 1978(a); Fujita *et al.* 1978(b); Suzuki *et al.* 1979; Yagi *et al.* 1982; Yagi *et al.* 1985).

The chemical structures of plant proteins have not been elucidated precisely because of the difficulty to purify them and the non-homogeneity of their function. Especially, it is very difficult to extract proteins from aloe because of the presence of phenolic compounds and the viscous mucopolysaccharides called mucin. In addition, it is technically difficult to separate proteins because most proteins present in aloe are presumably combined with sugars as glycoproteins. Therefore, it is not easy to characterize the exact chemical structure. Despite these aspects, the CAP project attempted to separate proteins contained in aloe based on previous research reports. Accordingly, several proteins from aloe were purified and partially characterized, including lectins, protease inhibitor, alprogen, and G1G1M1DI2, which is a glycopeptide fraction showing proliferation activity of epithelial cells. The toxicity (or side effects) of most synthetic drugs that have been developed or are being developed poses a problem for disease treatment in many cases. Therefore, systematical scientific approaches investigating the effective components of aloe may be meaningful and have clinical value due to its long history of therapeutic use in the civilian sector.

[1] School of Life Sciences and Biotechnology, Korea University
[2] Department of Chemistry, PuKyong National University

### 3.2.1. Aloe lectin

Lectins exert a variety of physiological functions such as agglutination and stimulation of mitotic cell division of lymphocytes through sugar-specific complex formation. They also have various physiological activities such as anti-ulcer and anti-tumor (Sharon and Lis 1972). Lectin recognizes and forms complexes with mono- or oligosaccharide molecules present on the surface of various cells such as red blood cells, fibroblasts or spermatozoa to combine in agglutination or clump formation. This enables the cells to recognize foreign cells and even to stimulate macrophages for endocytosis. In addition, the stimulation of non-dividing lymphocytes into the stage of growth and proliferation can be triggered by the specific binding of foreign antigens to surface receptors of the cells. Likewise, lectin acts as a mitogenic factor to stimulate lymphocyte proliferation and growth. In fact, this type of mitogenic effect can be observed from various cells, which indicates that glycoproteins or glycolipids on the cell surface may act as mediators of signal transduction during cell proliferation and differentiation (Sharon and Lis 1980, 1989, 1990; Barondes 1981).

Structural differences on the cell surface can be observed between cancer cells and normal cells, or during the development and differentiation of cells. In effect, changes in composition of sugar chains of glycoproteins or glycolipids or structural changes could be recognized by lectin to give rise to agglutination.

Several lectins are known to be toxic to mammalian cells. Especially, ricin, identified in the seed of caster oil plant, is a plant toxin and is a heterodimer composed of A-chain, which contains ribosome inactivation activity, and B-chain, which possesses agglutination property. Both abrin, which is purified from *Abrus precatorius*, and ricin were reported to show stronger anti-cancer activity, without significant side effect, than chemotherapeutic agents, especially in the case of breast cancer (Lord 1987).

For purification of lectin from *Aloe vera*, fresh aloe leaves were blended in a Waring blender and centrifuged at low RPM to remove any insoluble materials. The use of charcoal in the extraction removed the phenolic compounds and pigments effectively, which prevented the color change to dark brown. It has been identified that almost 60% of lectin activity remained in the fraction of 0-30% ammonium sulfate precipitation. However, it also contained viscous materials, presumably a lot of which was due to mucin-like carbohydrates, which made the purification of lectin difficult. Therefore, 55-80 % ammonium sulfate fraction was chosen for further purification of lectin to bypass this problem. Then, a series of column chromatographies was applied to purify the lectin such as DEAD anion exchange, Con A Sepharose and Sephacryl HR S-300 gel filtration. Finally, three peaks which express lectin activity were detected using a hemagglutination assay, as seen in Figs. 1 and 2. Because lectin was not purified as a single molecule until gel filtration column chromatography, the partially purified fraction was determined in terms of sugar specificity in the complex formation. As a result, hemagglutination activity of α-D-glucose was higher than that of fructose or N-acetyl glucosamine so that affinity resin was made by binding glucose to Sepharose CL-4B resin. Using this affinity column chromatography, the lectin was finally purified and its molecular weight was determined by passing through a Sephacryl S-100 HR gel filtration column with bovine serum albumin (BSA) and ovalbumin (OA) as controls. Its molecular weight was estimated as 55 KDa and two bands of 26 and 28 KDa in mass were detected on SDS-PAGE (Fig. 3). Therefore, the lectin purified from *Aloe vera* was identified as a glycoprotein consisted of two subunits as a heterodimer with a pH value of 5.1 (Chung and Park 1996).

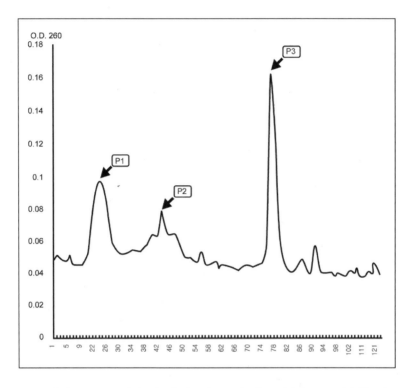

**Fig. 1.** The elution pattern of aloe lectin fraction on Sephacryl S-300 HR gel filtration. The aloe lectin fraction was prepared by the Con A Sepharose column eluted with ethylene glycol

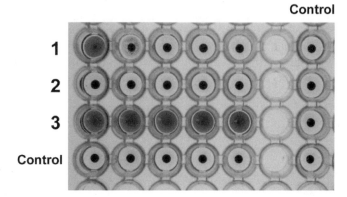

**Fig. 2.** Visual assay on microtiter plates of the hemagglutinating activity of peaks 1 to 3 purified on Sephacryl S-300 HR gel filtration, using rabbit erythrocytes. Each well contained 0.05ml of erythrocyte suspension and the first well contained 0.05ml of peak solution, with serial two-fold dilutions in adjoining wells. The control contained only rabbit erythrocyte suspension and 20 mM Tris·Cl, pH 7.4. (See Plate 1.)

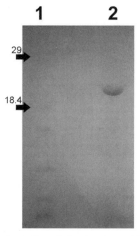

**Fig. 3.** The pattern of 16.5% SDS-PAGE of aloe lectin purified by glucose-affinity column chromatography. The arrows in Lanes 1 and 2 represent a size marker and aloe lectin, respectively. (See Plate 2.)

### 3.2.2. Glycoprotein Fraction: G1G1M1DI2

Aloe is known as an important traditional medicine in many countries and has been commonly used for health, medical, and cosmetic purposes (Blitz *et al.* 1963; Lee *et al.* 1980; Norton 1961). Some people keep one or a few plants at home to provide a readily available gel source for treatment of burns or other wounds. The chemical constituents of *Aloe* species have been investigated by several groups (Park *et al.* 1997; Reynolds 1985a; Reynolds 1985b), and the aromatic derivatives of the constituents have been reported to be effective in wound healing, antitumor, antimicrobial, immune modulating, peptic, laxative, and gastric juice controlling activities (Hirata and Suga 1977; Speranza *et al.* 1993).

As a part of the chemical studies on biologically active metabolites from aloe, we have investigated the bioassay-guided fractionation and isolation of bioactive compounds from a freeze-dried aloe gel.

The freeze-dried aloe gel used as raw material was obtained from Nam Yang Aloe Co. The aloe gel (25 g, designated as G1) was suspended in 200 ml of distilled water and 4 volumes of 95% EtOH were added. This solution was allowed to stand for 12 hr at 4.. The clear yellowish supernatant was decanted off from the white residue and the precipitate was centrifuged for 30 min at 4. (13,000 xg).

The supernatant was combined, evaporated, and lyophilized to give a mixture of small and medium size molecules (G1G1, 14.5 g). One gram of G1G1 was applied to an Amberlite XAD-2 column (250 ml of resin, 2.5×25 cm). The column was eluted with distilled water (500 ml), 50% aq. MeOH (200 ml), and MeOH-acetone (1:1, 400 ml), successively. The eluate with 50% aq. MeOH was concentrated and lyophilized to afford a dark-gray powder (G1G1M1, 53 mg), which was active on the thymidine uptake assay (Kopan, 1989).

G1G1M1 (53 mg) was dissolved in deionized water, then dialyzed for 3 days (m.w. 5,000 cut), and lyophilized (G1G1M1D, 23 mg). This fraction (G1G1M1D, 23 mg) was dissolved on 0.02 M $NH_4HCO_3$ and applied to an ion-exchange column (2.5×20 cm) of diethylaminoethyl (DEAE)-Toyopearl 650M.

The column was equilibrated and eluted with 0.02 M $NH_4HCO_3$ (600 ml), and then eluted with 0.3 M NaCl (600 ml). Fractions of 15 ml were collected and analyzed by UV (210 and 280 nm) and phenol-sulfuric acid method to give two fractions, G1G1M1DI1 and G1G1M1DI2, according to the elution pattern (Fig. 4). Each fraction was concentrated, dialyzed, and lyophilized. The yields of G1G1M1DI1 and G1G1M1DI2 were 7.8 mg and 2.7 mg, respectively.

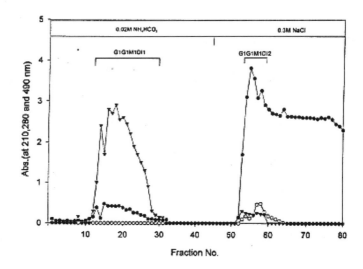

**Fig. 4.** Elution profile of G1G1M1D by DEAE-Toyopearl 650M CC. The sample was dissolved in 0.02 M $NH_4HCO_3$, and the column was eluted with 0.02 M $NH_4HCO_3$ and then with 0.3 M NaCl at 4□. Each fraction was 15ml. (•): absorbance at 210 nm; absorbance at 280 nm; absorbance at 490 nm (phenol-$H_2SO_4$ method).

Among them, G1G1M1DI1 did not show any activity in the thymidine uptake assay, but the main activity was found in the binding fraction G1G1M1DI2 (Table 1). G1G1M1DI2, a dark gray amorphous solid, was freely soluble in the water, but only slightly soluble in the alcohol and insoluble in the organic solvents. It was found to be homogeneous in the electrophoresis (Laemmli, 1970).

Elementary analysis (%) of G1G1M1DI2 indicated the following elemental composition: C, 35.8; H, 5.2; N, 2.5; and O, 56.5.

G1G1M1DI2 showed positive reactions with 1% $Ce(SO_4)_2$ in 10% aq. $H_2SO_4$ in TLC (silica gel) and with aniline phthalate in TLC (cellulose). It also revealed a band with periodic acid-Schiff's reagent (Dubray and Bezard, 1982; Kinchel and Bryan, 1990) and with Coomassie reagent (Bollag and Edelstein, 1991) in electrophoresis (20% SDS-PAGE), which are specific reagents to detect carbohydrate and protein, respectively.

The IR spectrum of G1G1M1DI2 also showed absorption bands at 3359, 1601, 1382, and 1077, indicating the presence of carbohydrate. Although the NMR spectra ($^1H$ and $^{13}C$) were poorly resolved, the $^{13}C$ NMR spectrum (100 MHz, $D_2O$) showed signals ascribable to aliphatic carbons ($\delta c$ 12.7, 12.9, 16.2, 21.4, 24.0, 32.8), carbonyl carbons ($\delta c$ 160.5, 171.8, 186.5), geminal carbons of oxygen function ($\delta c$ 61.2, 61.6, 62.0, 69.6, 70.6, 72.1, 75.7, 76.5), and anomeric carbons ($\delta c$ 100.9, 101.1). The $^1H$ NMR spectrum (400 MHz, $D_2O$) also showed signals assignable to carbohydrate moiety ($\delta$ 3.4-4.5, a mass of signals).

The evidence above has led us to assume a glycopeptide for G1G1M1DI2. Glycopeptide G1G1M1DI2 contained 32.6% of the total protein as egg albumin (Dubois *et al.* 1956) and 20.9% of the total carbohydrate as mannose (Lowry *et al.* 1951) (Table 1). G1G1M1DI2 had a molecular weight of 5,500 daltons, estimated by SDS gel electrophoresis. A peptide moiety of G1G1M1DI2 was obtained by degradation with $NaIO_4$, and the carbohydrate moiety by digestion with pronase E. The presence of each moiety was confirmed by electrophoresis. The carbohydrate moiety was suggested to contain fucose, galactose, glucose and mannose (Fig. 5), and the molar ratio (%) of these component sugars was 0.5, 2.4 48.8 and 48.3 (Table 2). To determine the amino acid composition and contents, the HCl-hydrolysate of the peptide moiety of G1G1M1DI2 was analyzed by amino acid analyzer. Fifteen kinds of amino acid were detected, and glutamic acid and glycine comprised 41.2 molar percent of the total detected amino acids (Table 3). From these observations, G1G1M1DI2 was characterized as a glycopeptide (Yang *et al.* 1998). The glycopeptide G1G1M1DI2 exhibited 6.45-fold more active than the control in thymidine uptake assay (Table 1), and also showed significant activities of wound healing in experiments using SCC 13 (squamous cell carcinoma 13) cells.

This result suggests that this glycopeptide has a skin cell proliferating activity. The detailed result and the molecular mechanism of wound healing have been ppublished elsewhere(Choi *et al.* 2001).

**Table 1.** Thymidine Uptake, and Contents of Carbohydrate and Protein in Active Fractions Purified from the Gel (**G1**) of *Aloe vera*

| Sample | Thymidine Uptake[a] | Carbohydrate[b] | Protein[b] |
|---|---|---|---|
| G1 | 210 | 47.9 | 2.6 |
| G1G1 | 270 | 50.7 | 3.9 |
| G1G1M1 | 320 | 62.6 | 11.8 |
| G1G1M1DI2 | 645 | 20.9 | 32.6 |

a) Relative activity(%) to the control, taken as 100%, in [$^3$H]thymidine uptake assay.
b) Presented as a percentage and determined by the phenol-sulfuric acid and Lowry's methods, for carbohydrate and protein, respectively.

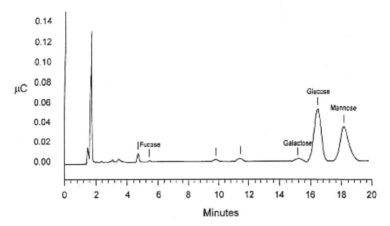

**Fig. 5.** HPLC Chromatogram of acid hydrolysate obtained from carbohydrate moiety of G1G1M1DI2. The column (CarboPak PA1, 4×250 mm) was eluted with $H_2O$-0.2 M NaOH (92 : 8) at room temperature and a flow rate of 1.0 ml/min.

**Table 2.** Monosaccharide Composition of **G1G1M1DI2**

| Monosaccharide | Composition (molar ratio, %) |
|---|---|
| fucose | 0.5 |
| galactose | 2.4 |
| glucose | 48.8 |
| mannose | 48.3 |

**Table 3.** Amino Acid Composition of G1G1M1DI2

| Amino acid | Composition(%) |
|---|---|
| Asp | 12.68 |
| Thr | 0.72 |
| Ser | 0.94 |
| Glu | 22.70 |
| Gly | 18.47 |
| Ala | 10.96 |
| Val | 8.75 |
| Met | 0.62 |
| Ile | 2.98 |
| Leu | 8.50 |
| Tyr | 1.70 |
| Phe | 2.47 |
| Lys | 5.24 |
| Arg | 2.16 |
| His | 1.18 |

### 3.2.3. Proteinase Inhibitors

Proteinase inhibitors of the serine family are present throughout the entire plant variety and are contained in high concentration within the tissues of the storage organ. One example of their various functions is the protection of the plants from insects by synthesizing proteinase inhibitors against, for instance, the digestive enzymes such as trypsin of insects so that they reduce the plant's value as consumable food in the event an insect infiltrates a plant. Namely, their production is regarded as a self-defense mechanism that suppresses the digestive enzymes of the infiltrating insect. In other more severe cases, the insect could be killed. Additionally, the trypsin inhibitor which is isolated from many plants has been reported to show anticancer activity in animal and *in vitro* experiments (Chem *et al.* 1992). In this study, the trypsin inhibitor reduced the formation of oxygen free radicals, suppressed cancers such as colon, rectal and lung cancers caused by chemicals, reduced spontaneous chromosome abnormality, and also prevented the infiltration and metastasis of cancer (Mignatti and Rifkin 1993).

An interesting feature of these inhibitors is their stability at high temperature in cooking and in highly acidic environments such as the stomach of humans and animals. In fact, people who consume foods containing these inhibitors are less likely to have cancers of the colon, breast, prostate, or skin than those who do not. In an experiment using mice, it has been identified that proteinase inhibitors of serine, aspartic acid or cysteine families are present in their stomach. The secretion of these inhibitors has been reported to influence the deterioration of the stomach mucosa (Nagy *et al.* 1987). On the other hand these inhibitors maintain a balance among each other in terms of activation, secretion and inhibition of their activities so that they protect the stomach by treating stomach ulcers. Moreover, proteinases of the serine family and their inhibitors may play an important role in preventing emphysema (Kerr *et al.* 1985), pneumonia (Welgus *et al.* 1987), arthritis (Baici and Lang 1990), and inflammatory large intestinal diseases (Granger *et al.* 1989). Considering the useful qualities of proteinase inhibitors of the serine family, attempts have been made to identify them in aloe.

The AVPI (*Aloe vera* proteinase inhibitor) has been obtained in the following procedure. Fresh leaves of *Aloe vera* were thoroughly blended with the Waring blender, and then underwent precipitation in a 25-40% ammonium sulfate. This fraction was placed in a boiling water bath for 10 min, and then underwent a series of column chromatographies such as Sephacryl S-100 HR, QAE-Sephadex A-25 anion -exchange, Phenyl-Sepharose CL-4B hydrophobic interaction and HPLC Protein Pak 125 gel filtration to obtain AVPI (Fig. 6, Cho    1997).

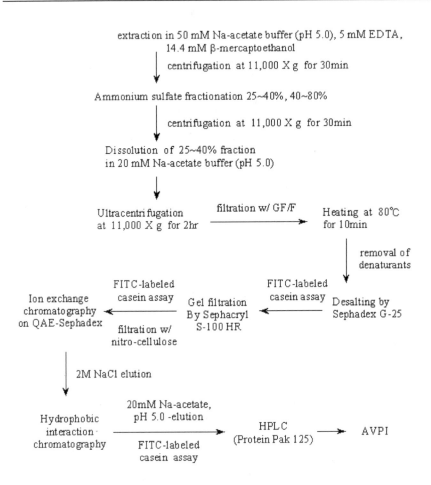

extraction in 50 mM Na-acetate buffer (pH 5.0), 5 mM EDTA,
14.4 mM β-mercaptoethanol

centrifugation at 11,000 X g for 30min

Ammonium sulfate fractionation 25~40%, 40~80%

centrifugation at 11,000 X g for 30min

Dissolution of 25~40% fraction
in 20 mM Na-acetate buffer (pH 5.0)

Ultracentrifugation          filtration w/ GF/F          Heating at 80℃
at 11,000 X g for 2hr    ————————————————→         for 10min

removal of
denaturants

Ion exchange       FITC-labeled              FITC-labeled
chromatography     casein assay    Gel filtration    casein assay    Desalting by
on QAE-Sephadex  ←——————    By Sephacryl   ←——————    Sephadex G-25
                   filtration w/     S-100 HR
                   nitro-cellulose

2M NaCl elution

                   20mM Na-acetate,
Hydrophobic        pH 5.0 -elution
interaction     ————————————————→    HPLC       ————→    AVPI
chromatography     FITC-labeled     (Protein Pak 125)
                   casein assay

**Fig. 6.** Purification procedure of *Aloe vera* proteinase inhibitor (AVPI)

Then the inhibition activity against trypsin was compared with fractions obtained by Phenyl-Sepharose CL4B column chromatography and soybean trypsin inhibitors using the FITC-labelled casein assay method. As a result, the inhibition activity of the fraction eluted with 20 mM Na-acetate buffer (pH 5.0) was 10 times greater than that of the soybean trypsin inhibitor (Table 4). This fraction was applied to 12.5% SDS-PAGE and a major band appeared at 16 KDa and a weak band at 10 KDa, as seen in Fig. 7.

The inhibition activity of proteinase was determined on the fractionated pattern of 12.5% SDS-PAGE by using reverse zymography (Fig. 8). Because the substrate (collagen) of trypsin is incorporated into the gel, the proteins fractionated in addition to the collagen on the gel could be degraded after the incubation in trypsin solution. If proteinase inhibitors are present, undigested protein bands after the trypsin treatment remain dark (Heussen and Dowdle 1980; Matsuzawa et al. 1996; Gomez et al. 1994). As seen in Fig. 8, bands of soybean trypsin inhibitor and AVPI (not clear band but smeared) showed dark bands by silver staining whereas the trypsin showed an unclear staining pattern. This result indicates that AVPI possesses the proteinase inhibition activity although the pattern of protein bands was smeared. The two protein bands detected on 12.5% SDS-PAGE could not be separated by conventional gel filtration chromatography, so that the HPLC method was applied for separation (Fig. 9). The two peaks shown in HPLC were examined for proteinase inhibition activity and the major peak (16 KDa) showed the inhibition activity.

This purified APVI showed a similar curve of inhibition as that of other enzyme inhibitors. Moreover, it showed 5% more inhibition activity against chymotrypsin, which is another member of serine proteinases, than against trypsin (Table 5). The APVI isolated from *Aloe vera* is considered to be a serine proteinase inhibitor, which is heat and acid stable and has the pH value of around 8.6.

**Table 4.** Results of FITC-labeled casein assay for fractions from hydrophobic interaction chromatography and soybean trypsin inhibitor

| Sample | 1M Ammonium sulfate-eluate (50 $\mu$g/ml) | 20 mM Na-acetate- eluate (50 $\mu$g/ml) | 50% Ethylene glycol- eluate (50 $\mu$g/ml) | Soybean trypsin inhibitor (500 $\mu$g/ml) |
|---|---|---|---|---|
| % Inhibition | 10.4 | 46.3 | 34.6 | 48.8 |

Negative control has 20 mM Na-acetate buffer added instead of inhibitor sample.

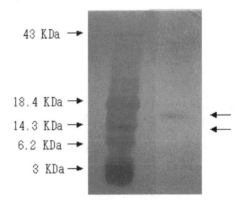

**Fig. 7.** The band pattern of *Aloe vera* proteinase inhibitor separated on 12.5% SDS-PAGE. (See Plate 2.)

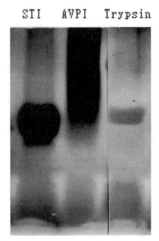

**Fig. 8.** Reverse zymography of AVPI (150 μg of Soybean trypsin inhibitor, 100 μg of *Aloe vera* proteinase inhibitor, 150 μg of Trypsin). (See Plate 3.)

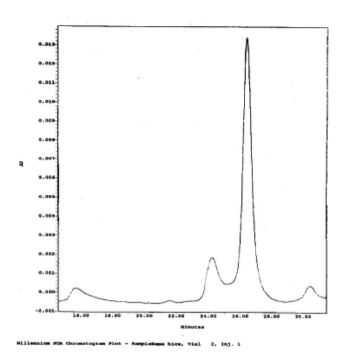

**Fig. 9.** Pattern of protein peaks of AVPI sample from HPLC results

**Table 5.** Comparison of inhibitory activity against chymotrypsin and trypsin

| Enzyme | Trypsin | Chymotrypsin |
|---|---|---|
| % inhibition | 80.2 | 83.9 |

Negative control has 20 mM Na-acetate buffer added instead of inhibitor sample.

### 3.2.4 Alprogen: antiallergic glycoprotein

Allergic diseases are found in 15~20% of the worldwide population. In Korea, the number of people with allergies is increasing rapidly. Also called the 'civilization disease', allergies are spreading rapidly throughout the world and pose a serious health problem for the 21st century. With the improvement of living standards, living quarters now maintain a steady temperature and moisture level, which contributes to the increased cultivation of fungi, mites and dust particles. In addition, the increasing use of synthetic fabrics and plastics such as nylons and polyesters generates dust which is produced during the process of their production. Along with pollens and animal hairs, they cause allergic reactions by invading the respiratory system. Other allergies include food allergies, allergies by contact with cosmetics, detergents or metals, anaphylaxis by drugs such as penicillins, and physical allergies by cold or hot temperatures, pressure or exposure to light. Most allergies show signs of coughing and skin rash, and cause respiratory problems and inflammatory diseases. In particular, asthma is considered to be one of the most dangerous allergic diseases. Allergies are immune diseases caused in high frequency by various factors such as temperature, moisture, region, season, environmental pollution, exercise and occupational work, and no ultimate remedy has yet been developed. Therefore, the development of an effective anti-allergic drug would benefit the welfare of our civilization, as well as have an enormously positive effect on our economy, industries and citizen well-being.

Most anti-allergic drugs for alleviation of allergic symptoms such as rash or itching are anti-histamines produced by organic synthesis, and, in severe cases, steroids, but none of them are known to be effective drugs against allergies. Terfenadine, the 3rd generation anti-histaminic drug domestically used in common against allergic symptoms such as skin problems, is now considered to cause circulation problems such as heart failure when used with antibiotics or antimicrobial agents. Steroids are also used to remove inflammation but their long-term side-effects cannot be ignored.

Most allergic symptoms are caused by mediators (histamines, leukotrienes, prostaglandins, PAF or TNF etc.) released from mast cells and/or basophils activated by antigen-antibody reactions. Nevertheless, it is difficult to extract a sufficient amount of basophils from the human or animal blood, and even more to obtain pure basophils. Therefore, mast cells are commonly used for screening the hypersensitivity reaction of allergy in terms of release of mediators upon activation of the cells, although there are functional differences between basophils and mast cells. In general, the anti-allergic effect has been examined to assay the inhibition of histamine release from mast cells of the mouse abdominal cavity. However, this system for the study of allergy is not considered effective for two reasons: the mouse system is not closely related to the human system, and mast cells of the abdominal cavity are known to be related with foods so that it is not proper to test against respiratory allergic symptoms through bronchus and lung. In addition, mast cells from different tissues show heterogeneity Furthermore, the inhibition of histamine release activity is currently being used for the development of anti-allergic drugs, and the inhibition of leukotrienes release also warrants research attention because it is 1,000 times more potent than histamine.

Therefore, the Guinea pig has been chosen as a model animal because of its similarity in hypersensitivity reaction to humans. A glycoprotein which inhibits the release of histamine, leukotrienes and/or TNF has been successfully purified from aloe by using Guinea pig lung mast cell, and was named "alprogen".

Ten kilograms of fresh aloe leaves cultivated locally were mixed in a Waring blender in 20 liters of extraction buffer containing 0.1 M Na-phosphate buffer, pH 8.0, 14.4 mM β-mercaptoethanol, 1% PVP (polyvinylpyrrolidone, 1 mM EDTA). After removal of insoluble substances at low speed centrifugation, the resulting supernatant was then precipitated with 25-55% $(NH_4)_2SO_4$ to remove viscous materials, mainly carbohydrates. Ten grams of proteins were obtained by passing through Sephadex G-25 column chromatography. Successive purifications with DEAE-Sephacel anion exchange and Phenyl-Sepharose CL-4B hydrophobic interaction column chromatographies produced a partially purified active protein fraction which was named NY945 (Ro *et al.* 1998). Especially, Phenyl-Sepharose column chromatography removed most of the non-active proteins. The pure single compound of alprogen was finally purified by Mono Q ion exchange column chromatography and Superdex 75 gel filtration column chromatography (Fig.'s 10 and 11, Table 6, Ro *et al.* 2000; Park *et al.* 2000). Its molecular weight was identified as 7 KDa on 16.5% SDS-PAGE (Fig. 12). However, two peaks obtained by DEAE-Sephacel, anion exchange, column chromatography were separately purified with successive column chromatographies with ConA Sepharose affinity, Phenyl-Sepharose CL-4B hydrophobic interaction and Superdex 75 gel filtration and they were named as alprogen I and II, respectively (Table 7). Alprogen I was identified to contain two subunits of 8 and 12 KDa in molecular weight on 16.5% SDS-PAGE while alprogen II was also found to consist of two different subunits of 7 and 8 KDa in molecular weight. Thus, both alprogen I and II are present as heterodimers. However, the molecular weights of these two alprogens were measured as 11.93 and 11.92 KDa, respectively, which was different from their estimation on gel electrophoresis. This difference might be explained by the presence of sugars on the polypeptide chain which influences the migration rate on SDS-PAGE (Lee 1998).

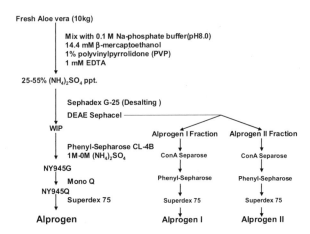

**Fig. 10.** Procedure for purification of Alprogens from *Aloe vera*

**Table 6.** Summary of the purification procedure of Alprogen from *Aloe vera*

| Procedure | Total protein (mg)[b] | Total activity (U)[c] | IC$_{50}$[d] | Purification (-fold) | Recovery (%) |
|---|---|---|---|---|---|
| 25-55% (NH$_4$)$_2$SO$_4$ ppt. | N.D. | | | | |
| Sephadex G-25 | 10,000 | 10,000 | 1mg | | 100 |
| DEAE Sephacel (W1P) | 160 | 1,600 | 100 $\mu$g | 10 | 16 |
| Phenyl Sepharose | 30 | 1,000 | 30 $\mu$g | 33 | 10 |
| Mono Q | 2 | 400 | 5 $\mu$g | 198 | 4 |
| Superdex 75 | 0.4 | 200 | 2 $\mu$g | 396 | 2 |

a) Mast cells were purified by enzyme digestion and a discontinuous Percoll density gradient centrifugation. Mast cells ($0.4 \times 10^6$) sensitized by anti-OVA were challenged by OVA (0.1 $\mu$g/m$\ell$). Proteins of each purification procedure were added 5 min before antigen challenge.

b) Protein concentration was determined by Bradford method.

c) Unit is the amount of protein that inhibits 50% of histamine release in the mast cells ($0.4 \times 10^6$) activated by OVA and anti-OVA complexes, compared to that of OVA challenge alone.

d) The value of IC$_{50}$[c] was represented as an amount of protein able to inhibit 50% of histamine released in a solution of mast cells ($0.4 \times 10^6$) activated by OVA and anti-OVA complexes, compared to that of OVA challenge alone.

**Table 7.** Summary of the purification procedure of Alprogen I, II from *A. vera*[a]

| Procedure | Total protein (mg)[b] | Total activity (U)[c] | $IC_{50}$[d] | Purification (-fold) | Recovery (%) |
|---|---|---|---|---|---|
| 25-55% $(NH_4)_2SO_4$ ppt. | N.D. | | | | |
| Sephadex G-25 | 10,000 | 10,000 | 1mg | | 100 |
| DEAE Sephacel | 800 | 8,000 | 100 $\mu g$ | 12.5 | 80 |
| Phenyl Sepharose | | | | | |
| Alprogen I | 0.4 | 200 | 2 $\mu g$ | 2472 | 2.0 |
| Alprogen II | 1.2 | 600 | 2 $\mu g$ | 2472 | 6.0 |
| Superdex 75 | | | | | |
| Alprogen I | 0.24 | 160 | 1.5 $\mu g$ | 6170 | 1.6 |
| Alprogen II | 0.54 | 360 | 1.5 $\mu g$ | 6170 | 3.6 |

a) Mast cells were purified by enzyme digestion and a discontinuous Percoll density gradient centrifugation. Mast cells ($0.4\times10^6$) sensitized by anti-OVA were challenged by OVA (0.1 $\mu g/m\ell$). Proteins of each purification procedure were added 5 min before antigen challenge.
b) Protein concentration was determined by Bradford method.
c) Unit is the amount of protein that inhibits 50% of histamine release in the mast cells ($0.4\times10^6$) activated by OVA and anti-OVA complexes, compared to that of OVA challenge alone.
d) The value of $IC_{50}$[c] was represented as an amount of protein able to inhibit 50% of histamine released in a solution of mast cells ($0.4\times10^6$) activated by OVA and anti-OVA complexes, compared to that of OVA challenge alone.

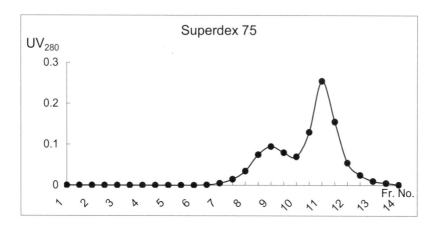

**Fig. 11.** A gel filtration chromatography profile of Alprogen on Superdex 75 10/30. The estimated volume of Alprogen is 10.1 ml. The arrow indicates the alprogen peak. (See Plate 3.)

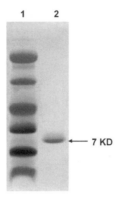

**Fig. 12.** Protein analysis of Alprogen by 16.5% SDS-PAGE. The gel electrophoresis was carried out through 16.5%(w/v) polyacrylamide gel containing 0.1% SDS. Lane 1 is a size marker, Lane 2 is Alprogen. (See Plate 4.)

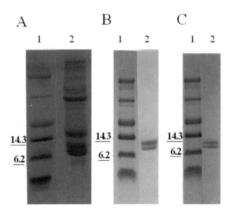

**Fig. 13.** Protein analysis of the NY945 and Alprogens by 16.5% SDS-PAGE. The gel electrophoresis was carried out through 16.5% (w/v) polyacrylamide gel containing 0.1% SDS. A: lane 1, protein size marker; lane 2, NY945. B: lane 1, protein size marker; lane 2, Alprogen I. C: lane 1, protein size marker; lane 2, Alprogen II. (See Plate 4.)

**References**
1. Ashely FL, O'Loughin BJ, Peterson R, Fernandez L, Stein H, Schwartz AN (1950) The use of *Aloe vera* in the treatment of thermal and irradiation burns in laboratory animals and humans Plastic and Reconstructive Surgery 20, 383-396
2. Baici A, Lang A (1990)    Cathepsin B Secretion by Rabbit Articular Chondrocyte Modulation by Cycloheximide and Glycosaminoglycan Cell Tissue Res 259, 567-573
3. Barondes SH (1981) Lectins Their multiple endogenous cellular functions. Ann Rev Biochem 50, 207-231
4. Bigas MR, Cruz N, Suarez A (1988) Comparative evaluation of aloe vera in the management wounds in Guinea pigs Plastic and Reconstructive Surgery 81(3), 386-389
5. Blitz JJ, Smith JW, Gerard JR (1963) Aloe vera gel in peptic ulcer therapy Preliminary report J of American Osteopathic Association 76(2), 61-66
6. Blumenkrantz N, Asboe HG (1973) New method for quantitative determination of uronic acid Anal Biochem 54, 484-489
7. Bollag DM, Edelstein SJ (1991) Gel electrophoresis under denaturing condition Protein Methods Wiley-Liss Inc New York pp.95
8. Campbell EJ, Senior RM, Welgus HG (1987)  Extracellular Matrix Injury During Lung Inflammation Chest 92 161-169
9. Cera LM, Heggers JP, Hagstrom WJ, Rodson MC (1980) The therapeutic efficacy of *Aloe vera* cream in thermal injuries two case reports. Journal of American Animal Hospital Association 16 768-772
10. Chen P, Rose J, Love R, Wei CH, Wang B (1992) -C. Reactive Sites of an Anticarcinogenic Bowman-Birk Proteinase Inhibitor Are Similar to Other Trypsin Inhibitors. Biol Chem 267 1990-1994
11. Cheney RH (1970) Aloe drug in human therapy QJ Crude Drug Res 10, 1523-1529
12. Cho HJ (1997) Purification and Characterization of Serine Proteinase Inhibitor from Aloe vera Master's Thesis Korea University
13. Choi SW, Son BW, Son YS, Park YI, Lee SK, Chung MH (2001) The wound-healing effect of a glycoprotein fraction isolated from aloe vera Brit J of Dermatol 145, 535-545
14. Chung YJ, Park YI (1996) Purification and partial characterization of lectin from Aloe vera J Inst Biotechnol Korea University 8, 71-78
15. Cole HN, Chen KK (1943) Aloe vera in oriental dermatology Arch. Dermatol Syph 47, 250
16. Collin CE, Collin C (1935) Roentgen dermatitis treated with fresh whole leaf of Aloe vera. Am J of Roentgenology 33, 396-397
17. Crosswhite FS (1984) Aloe vera, plant symbolism and the threshing floor Desert Plant 6 43-50
18. Davis RH, Agnew PS, Shapiro E (1986) Antiarthritic activity of anthraquinones found in aloe for podiatric medicine J of Am Podiatric Med Ass 76(2) 61-66

19. Davis RH, Agnew PS, Shapiro E (1987a)    Aloe vera and wound healing J of Am Podiatric Med Ass 77(4) 165-169

20. Davis RH, Agnew PS, Shapiro E (1987b) Topical anti-inflammatory activity of aloe vera as measured by ear swelling J of Am Podiatric Med Ass 77(11), 610-612

21. Davis RH, Agnew PS, Shapiro E (1988) Aloe vera a natural approach for treating wounds edema and pain in diabetics J of Am Podiatric Med Ass 78(2), 60-68

22. Davis RH, Agnew PS, Shapiro E (1989) Wound healing, oral and topical activity of aloe vera J of Am Podiatric Med Ass 79(11), 559-562

23. Dubois M, Gilles KA, Hamilton JK, Reberts PA, Smith F (1956) Colorimetric method for determination of sugars and related substances Anal Chem 28 350-356

24. Dubray G, Bezard G (1982) A highly sensitive periodic acid-silver stain for 1,2-diol groups of glycoproteins and polysaccharides in polyacrylamide gels Anal Biochem 119 325-329

25. Fly LB, Keim I (1963) Tests of aloe vera for antibiotic activity Econ Botany 17 46-48

26. Fujita K, Ito S, Teradaira R, Beppu H (1978a)    Property of a carboxypeptidase Biochem Pharma 28 1261-1262

27. Fujita K, Suziki I, Ochiai J, Shinpo K, Inoue S, Satio H (1978b) Specific reaction of aloe extract with serum proteins of various animals Experientia 34, 523-524

28. Fujita K, Teradaira R, Nagatisu T (1976) Bradykininase activity of aloe extract Biochem Pharma 25, 205

29. Grindllay D, Reynolds T (1986) The aloe vera phenomenon : a review of the properties and modern uses of the leaf parenchyma gel J Ethnopharmacol 16, 117-151

30. Kinchel CO, Bryan A (1990) Detection of glycoproteins separated by nondenaturing polyacrylamide gel electrophoresis using the periodic acid-Schiff stain Anal Biochem 187, 147-150

31. Laemmli U K (1970) Nature 227, 680

32. Lee BC (1998) Purification and characterization of the alprogens from Aloe vera: Novel glycoproteins with inhibitory effects on type I hypersensitivity reaction Ph.D. thesis Korea University.

33. Lord JM (1987)    The use of cytotoxic plant lectins in cancer therapy Plant Physiol 85, 1-3

34. Lorenzetti LJ, Salisburg R, Beal J (1964)    Bacteriostatic property of aloe vera J Pharm Sci 53, 1287

35. Lowry OH. Rosebrough NJ, Farr AL, Randall RJ (1951) Protein measurement with the Folin-phenol reagent J Biol Chem 193, 265-275

36. Mignatti P, Rifkin DB (1993) Biology and Biochemistry of Proteinases in Tumor Invasion Physiological Reviews 73, 161-194.

37. Morton JF (1961) Folk uses and commercial exploitation of the aloe leaf pulp Econ Botany 15, 311-317

38. Nagy L, Johnson BR, Hauschika P, Szabo S (1997) Characterization of Proteases and Protease Inhibitors in Rat Stomach Am J Physiol 272, G1151-G1158
39. Park YI, Kim JY, Chung YJ, Ro JY (2000) The inhibitory mechanism of alprogen Iiα on the mediator release in the mast cells sensitized with house dust mite antigen J Asthma Allergy Clin Immunol 20, 943-959
40. Reynolds T (1985) The compounds in aloe leaf exudates : a review Botanical J of the Linnean Society 90, 157-177
41. von Ritter C, Be R, Granger DN (1989) Neutrophilic Proteases: Mediators of Formyl-Methyonyl-Leucyl Phenylalanine-Induced Ileitis in Rats Gastroenterology 97, 605-609
42. Riley DJ, Kerr JS (1985)    Oxidant Injury of The Extracellular Matrix Potential Role in The Pathogenesis of Pulmonary Emphysema Lung 163, 1-13
43. Ro JY, Lee BC, Chung MH, Lee SK, Sung CK, Kim KH, Park YI (1998) The inhibitory mechanism of aloe glycoprotein (NY945) on the mediator release in the Guinea Pig lung mast cell activated with antigen-antibody complexes. Kor. J. Physiol. Pharmacol. 2, 119-131
44. Ro JY, Lee BC, Kim JY, Chung YJ, Chung MH, Lee SK, Jo TH, Kim KH, Park YI (2000) Inhibitory mechanism of aloe single component (alprogen) on mediator release in Guinea Pig lung mast cells activated with specific antigen-antibody reactions J Pharmacol and Expt Therapeutics 292(1),114-121
45. Saito H, Ishiguro T, Imanish K, Suzuki I (1982) Pharmacological studies on a plant lectin, Aloctin A. ... Inhibitory effect of aloctin A on experimental models of inflammation in rats Jap J Pharmacol 32, 139-142
46. Sasaki T, Uchida H, Uchida NA, Tokasuca N, Tachibana Y, Nakamichi K, Endo Y, Kamiya H (1987) Antitumor activity and immunomodulatory effect of glycoprotein fraction from scallop Patinopecten yessoensis Nippon Suisan Gakkaishi 53(2), 267-272
47. Sharon N, Lis L (1972) Lectins Cell-agglutinating and sugar-specific proteins Science 177, 949-959
48. Sharon N, Lis L (1980) Lectins in higher plants in The biochemistry of plants Academic press New York 371-447
49. Sharon N, Lis L (1989) Lectins as cell recognition molecules Science 246 227-246
50. Sharon N, Lis L (1990) Legume lectins- a large family of homologous proteins FASEB J 4 3198-3208
51. Suzuki I., Saito H, Inoue S, Migita S, Tagahashi, T (1979) Purification and characterization of two lectins from Aloe arborescens Mill J Biochem 85, 163-171
52. Takagi K, Ocabe S, Saziki R (1970) A new method for the production of chronic gastric ulcer in rats and the effect of several drugs on the healing J Pharmacol 19, 418-426
53. Twining SS (1984) Fluorescein Isothiocyanate-Labeled Casein Assay for Proteolytic Enzymes Anal Biochem143, 30-34

54. Yang MR, Kang CG, Roh YS, Son BW, Choi HD, Park YI., Lee SK, Choi SW, Chung MH (1998) The glycopeptide a promoter of thymidine uptake from Aloe vera Nat Prod. Sci 4(2), 62-67

55. Yagi A, Harada N, Yamada H, Iwadare S, Nishioka I (1982) Antibradykinin active material in Aloe saponaria J Pharm Sci 71, 1172-1174

56. Yagi A, Machii K, Nishimura H, Shida T, Nishioka I (1985) Effect of aloe lectin on deoxyribonucleic acid synthesis in baby hamster kidney cells. Experientia, 41, 669-671

57. Yamamoto I (1970) A new substance, aloe ulcin its chemical properties and inhibition on histamine synthetic enzyme J Med Soc Toho 17, 361-364

58. Yamamoto I (1973) Aloe ulcin a new principle of *Cape aloe* and gastro-intestinal function especially experimental ulcer in rats J Med Soc Toho 20, 342-347

59. Zawahry M, Hegazy MR, Helal M (1973) Use of aloe in treating leg ulcers and dermatoses Int'l J of Dermatol 12, 68-73

## 3.3.    Carbohydrates

**Kim, Young Shik, Ph.D.**

Natural Products Research Institute/College of Pharmacy,
Seoul National University

### 3.3.1. Introduction

Aloe, a member of the Liliaceae family, has been used medicinally for thousands of years. Of the several hundred species, *Aloe vera* Linne is the most famous and has been widely applied for commercial purposes as well as therapeutic effects. (Ang Lee 2003; Grindlay and Reynolds 1986; Reynolds and Dweck 1999) Aloe gel, obtained from the inner parenchymal tissue of the leaf, has been used for topical wound healing despite inconsistent results in clinical trials. In addition to topical administration, aloe is often used in oral dosage forms. It has been claimed to heal gastrointestinal ulcers, treat AIDS, lower blood sugar, treat cancer, and lower blood lipid levels. (Reynolds and Dweck 1999) Although aloe is potentially effective in some of these conditions, more clinical trials are required.

Aloe extract contains many compounds, some of which have not been characterized. Of them, polysaccharides and lectins are thought to be the most important components in the gel. They are responsible for its anti-inflammation and immunomodulating activities. The major polysaccharide of aloe gel is reported as acetylated mannan (Acemannan, ACM), which consists of a polydispersed $\beta$-1,4-linked mannan substituted with *O*-acetyl groups. Research interest in ACM has grown since the demonstration of its antiviral and antitumor properties *in vivo*. Parenteral administration of ACM has been followed by a greater regression of tumors such as fibrosarcomas in mice, dogs and cats. (King *et al.* 1995)

### 3.3.2. Chemistry of Aloe Polysaccharides

Since various biological properties have been demonstrated *in vitro* and *in vivo*, many reports have studied the structure of its active polysaccharide fractions. Despite the ongoing research, there is considerable discrepancy with respect to the structure of the polysaccharide isolate from *Aloe vera*.

Gowda *et al.* (Gowda *et al.* 1979) were the first to find that the polysaccharide isolated from gel by alcohol precipitation contained Man and Glc in a stoichiometric ratio of ~19:1. They suggested that the backbone of the structure is composed of *O*-acetylated →4)-$\beta$-Man-(1→ residues, with randomly substituted →4)-$\beta$-Glc-(1→   residues (Manna and McAnalley 1993). Manna and McAnalley determined the position of the *O*-acetyl group in a $\beta$-1,4-mannan (Manna and McAnalley 1993). Others have reported a structure containing Gal residue linked to *O*-6 of backbone →4)-$\beta$-Man-(1→ residues (Pugh *et al.* 2001). Pugh *et al.*

characterized a high-molecular weight polysaccharide from the same species with potent immunomodulatory activity (King et al. 1995). However, the polysaccharide was composed of diverse neutral sugars, including Glc, Gal, Man, Ara, Rham, and GlcA, which is quite different from the previous report.

Very recently, the polysaccharide isolated by alcohol precipitation of Aloe vera gel was reported to have a Man:Glc:Gal:GalA:Fuc:Ara:Xyl ratio of 120:9:6:3:2: 2:1, along with traces of Rha and GlcA (Tai Nin Chow et al. 2005). NMR analysis of oligosaccharides generated by endo-(1→4)-β-D-mannanase and acid hydrolysis showed the presence of di-, tri-, and tetra-saccharides of 4-β-Man, β-Glc-(1→4)-Man, β-Glc-(1→4)-β-Man-(1→4)-Man, and β-Man-(1→4)-[α-Gal-(1→6)]-Man, consistent with a backbone containing alternating →4)-β-Man-(1→ and →4)-β-Glc-(1→ residues in a ～15:1 ratio (Tai Nin Chow et al. 2005).

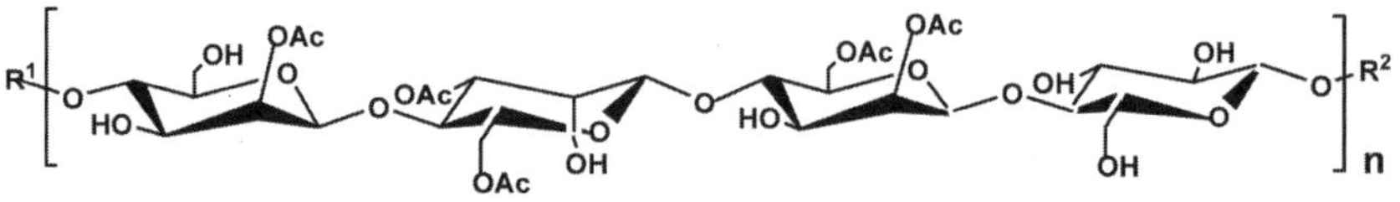

Fig. 1. A general structure of acetylated mannan from Aloe vera

These differences are speculated to arise from the degradation or contamination occurring during processing. In fact, there are many specific steps where polysaccharides can be lost during processing (Diehl and GmbH 1998).

Aloe contains, as do most plants, cellulase in its tissues. This enzyme can start to break down the major polysaccharide as soon as harvest and processing occur. Bacteria can also produce enzymes that break down polysaccharide. If bacteria grow in an uncontrolled fashion they accelerate the loss of polysaccharide. The same factors that help control the loss of polysaccharide from plant enzymes - low temperature and rapid processing - help minimize bacterial growth. As an example, the processed aloe gel powder was fractionated according to the molecular size using ultrafiltration. The molecular weight of the finally purified acemannan fraction (GC4) was over 3,000 Da. (Lee et al. 2001). Monosaccharide compositional analysis of the purified acemannan using gas chromatography indicated that it was composed of 97% mannose and 3% glucose. Acetylation was confirmed by IR spectrum. Stretching bands of carbonyl groups were clearly shown at 1236 and 1736 cm$^{-1}$, but they disappeared after alkaline hydrolysis. The multiple peaks in the NMR spectrum demonstrated that the molecule contained acetyl groups. In other cases, the yield of acemannan from the lyophilized gel of Aloe vera was about 2%. The average molecular weight of the purified acemannan was larger than 500,000 Da (unpublished data).

The major Aloe polysaccharide is important for two reasons. First is the biological activity of ACM, particularly with regard to the skin. This molecular appears to be an excellent emollient with important moisturizing capability. Furthermore, there are scientific suggestions that the polysaccharide enhances the ability of therapeutic agents to penetrate the skin, thus potentiating their beneficial effect.

Second, the level of polysaccharides is an excellent indicator of the general quality of the Aloe. An educated consumer will question any product that has been subjected to excessive and harsh processing. Low levels of polysaccharide indicate an Aloe preparation that has been roughly handled. Therefore, it should be readily apparent that good quality Aloe should have the highest possible level of polysaccharides (Pelley 1997).

### 3.3.3. Aloe polysaccharide and immunomodulation

#### 3.3.3.1. Aloe and Wound healing

Wound healing involves a complex series of interaction between different cell types, cytokine mediators, and the extracellular matrix (MacKay and Miller 2003). *Aloe vera* has been used for decades as a folk remedy for burns, wounds, and scars. Topical application and oral administration of *Aloe vera* to rats in treatment to heal dermal wounds increased the collagen content of the granulation tissue as well as the degree of cross-linkage. Wound repair is a complex series of coordinated events regulated by a delicately orchestrated cascade of cytokines and growth factors that restore the structural integrity of damaged tissue. Manipulation of the growth factor profile or wound environment through topical application of therapeutic agents could positively influence the rate and quality of wound repair. Transforming growth factor-beta, platelet-rich plasma, activated macrophage supernatant, and growth hormone are sources of mediators that may facilitate wound healing.

#### 3.3.3.2. Aloe and immunomodulation

The immune system provides the defense mechanisms of the body. It is concerned with defense against foreign cells and foreign substances. The most important cells are the types of white blood cells (leukocytes) and some special plasma proteins called antibodies. The white blood cells, known as lymphocytes, actually travel extensively within the body. Many of them, originating in the bone marrow, travel to the thymus gland, where their further development is influenced, and they then establish themselves in a number of centers around the body, especially the lymph glands in the neck, armpits, groin areas and spleen. Here they constitute centers of lymphoid tissue, as do the tonsils and the appendix which hence should be regarded as part of the immune system (Roitt *et al.* 2001).

Two other very important types of cell in the body's defenses, both of which are also white blood cells, are the macrophage and the neutrophil. Both of these cell types carry out the process called phagocytosis, which plays an important part in the overall processes of immunity. The actual phagocytosis step is really a cleaning up operation after some of the earlier immune processes have taken place. The phagocytosed item is neutralized and ends up being destroyed and eliminated. The digestive and oxidizing processes that take place within the phagocyte destroy the structure of the offending item and render it unrecognizable from its original form. The effect is therefore both protective and cleansing.

Obviously, anything which can make the process of phagocytosis more effective and more active is going to be significant for the processes of immunity.

*Aloe vera* polysaccharides between 400 and 5 kDa exhibit the most potent macrophage-activating activity as determined by increased cytokine production, nitric oxide release, surface molecule expression, and phagocytic activity. In accordance with the *in vitro* activity, polysaccharides between 400 and 5 kDa also exhibited the most potent antitumor activity in vivo (Im *et al.* 2005).

The immunomodulatory activity of acemannan on dendritic cells (DCs), which are the most important accessory cells for the initiation of primary immune responses, has been studied (Lee *et al.* 2001). Phenotypic analysis for the expression of class II MHC molecules and major co-stimulatory molecules such as B7-1, B7-2, CD40 and CD54 confirmed that acemannan could induce maturation of immature DCs. Functional maturation of immature DCs was supported by increased allogeneic, mixed lymphocyte reaction (MLR) and IL-12 production. The adjuvant activity of acemannan is at least in part due to its capacity to promote differentiation of immature DCs.

The high molecular weight fraction was further separated by gel filtration into two polysaccharide components, B1 (320, 000 Da) and B2 (200, 000 Da), largely composed of mannose (t'Hart *et al.* 1989). Both substances showed anti-complement activity at the C3 activation step. In one study, carcinogenesis by DNA-adduct formation was shown to be inhibited by a polysaccharide-rich, aloe fraction in vitro (Kim *et al.* 1999).

### 3.3.4. Aloe and hematologic activity

Studies using a pure (>99%) carbohydrate fraction from *Aloe vera* extracts revealed increased hematopoietic and hematologic activity compared to the starting material. In addition, this fraction differentially regulated liver and lung cytokine mRNA levels, thereby significantly increasing the message for hematopoietic cytokines (Talmadge *et al.* 2004), which is associated with an increase of G-CSF and SCF mRNA levels in the lung and the liver.

### 3.3.5. Metabolism of Aloe polysaccharide

The biodisposition of fluoresceinylisothiocyanate (FITC)-labeled aloemannan (FITC-AM) with the homogenate from some mice organs was demonstrated, and FITC-AM was metabolized to a smaller molecule (MW 3000) by the large intestinal microflora in feces. The aloe polysaccharide (MW: 80000), modified with cellulase under restricted conditions, immunologically stimulated the recovery of UV-B-induced tissue injury. Thus the modified polysaccharides of aloemannan, together with acemannan (MW: about 600,000), are expected to participate in biological activity following oral administration (Yagi *et al.* 1999).

**References**
1. Ang Lee, MK (2003) Most commonly used herbal medicine in the USA Parthenon Publishing New York 7-32.
2. Diehl B (1998) GmbH, R. Aloe vera. quality inspection and identification Agro-Food-Industry Hi-Tech 14-17
3. Gowda DC (1979) Neelisiddaiah BAnjaneyalu YV Carbohydr Res 72, 201-205.
4. Grindlay D, Reynolds T (1986) The Aloe vera phenomenon a review of the properties and modern uses of the leaf parenchyma gel J Ethnopharmacol 16, 117-51.
5. Im SA, Oh ST, Song S, Kim MR, Kim DS, Woo SS, Jo TH, Park YI, Lee CK (2005) Identification of optimal molecular size of modified Aloe polysaccharides with maximum immunomodulatory activity. Int Immunopharmacol : 5, 271-9.
6. Kim HS, Kacew S, Lee BM (1999) In vitro chemopreventive effects of plant polysaccharides (Aloe barbadensis miller Lentinus edodes Ganoderma lucidum and Coriolus versicolor) Carcinogenesis 20, 1637-40.
7. King GK, Yates KM, Greenlee PG, Pierce KR, Ford CR, McAnalley BH, Tizard IR (1995) The effect of Acemannan Immunostimulant in combination with surgery and radiation therapy on spontaneous canine and feline fibrosarcomas J Am Anim Hosp Assoc 1995, 31, 439-47.
8. Lee JK, Lee MK, Yun YP, Kim Y, Kim JS, Kim YS, Kim K, Han SS, Lee CK (2001) Acemannan purified from Aloe vera induces phenotypic and functional maturation of immature dendritic cells Int Immunopharmacol 1, 1275-84.
9. MacKay D, Miller AL (2003) Nutritional support for wound healing Altern Med Rev 8, 359-77.
10. Manna S, McAnalley BH (1993) Determination of the position of the O-acetyl group in a beta-(1-->4)-mannan (acemannan) from Aloe barbadensis Miller Carbohydr Res 241, 317-9.
11. Pelley PR (1997) The Story Of Aloe Polysaccharides IASC newsletter January
12. Pugh N, Ross SA, ElSohly MA, Pasco DS (2001) Characterization of Aloeride a new high-molecular-weight polysaccharide from Aloe vera with potent immunostimulatory activity J Agric Food Chem 49, 1030-4.
13. Reynolds T, Dweck AC (1999) Aloe vera leaf gel a review update J Ethnopharmacol 68, 3-37.
14. Roitt I, Brostoff J, Male D (2001) Immunology Mosby Edinburgh 1-12.
15. Tai Nin Chow J, Williamson DA, Yates KM, Goux WJ (2005) Chemical characterization of the immunomodulating polysaccharide of Aloe vera L Carbohydr Res 340, 1131-42.
16. Talmadge J, Chavez J, Jacobs L, Munger C, Chinnah T, Chow JT Williamson D, Yates K (2004) Fractionation of Aloe vera L inner gel purification and molecular profiling of activity Int Immunopharmacol 4, 1757-73.
17. t'Hart LA, van den Berg AJ, Kuis L, van Dijk H, Labadie RP (1989) An

anti-complementary polysaccharide with immunological adjuvant activity from the leaf parenchyma gel of Aloe vera Planta Med, 55, 509-12.

18. Yagi A, Nakamori J, Yamada T, Iwase H, Tanaka T, Kaneo Y, Qiu J, Orndorff S (1999) In vivo metabolism of aloemannan Planta Med 65, 417-20.

# 4. Efficacy of Aloe

*Aloe vera* (*Aloe barbadensis* Miller) is a perennial plant classified in the family of Liliaceae and contains thick and pulpy leaves. It looks like a cactus in shape but it is not related with a cactus at all. It grows well in a hot and dry climate, although in this environment aloe can be attacked by insects and damaged by UV, etc.This kind of extreme situation presumably encouraged aloe to develop various physiological activities such as wound healing effect, antimicrobial activity and so on.Historically, aloe has been used as a curing agent based on human experience rather than scientific evidence. Therefore, it is necessary to identify the exact biochemical mechanism of action employed by the physiologically active components present in aloe.Although there have been many attempts to identify the functions of aloe components, the CAP project has been organized systematically to elucidate various functions of any components present in *Aloe vera*.

## 4.1. Wound healing effect

## Chung, Myung Hee Ph.D. and Choi, Seong Won Ph.D.

College of Medicine, Seoul National University

### *4.1.1. The wound healing effect*

#### 1. Histology of epidermis

The epidermis is derived from ectoderm. It consists predominantly of stratified squamous cells in small numbers of Melanocytes, Langerhans and Merkel cells. The superficial surface is flat but the deep surface is undulating due to papillary folds extending into the upper dermis. The keratinocytes are arranged in four layers: basal, squamous, granular, and cornified. The dermis is organized in two zones: the upper (papillary zone) and lower (reticular zone).

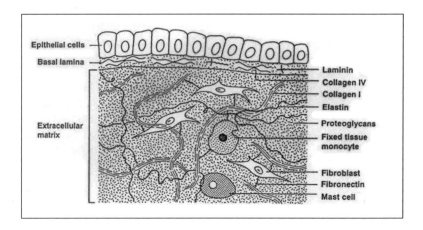

**Fig. 1.** The structure of the skin

## 2. Mechanism of wound healing

Wound healing proceeds through the steps of coagulation, inflammation, fibroblast formation, and matrix remodeling (Fig. 2 ~ 5). These reactions require various cytokines (Table 1).

### (1) Coagulation

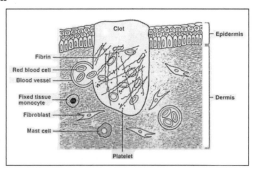

**Fig. 2.** Coagulation

As the blood vessel leaks, intravascular components move into the injury. Platelets attach to the damaged endothelial cells and to the fibrin network, which is stimulated by thromboxane A2. Platelets produce coagulation factors and cytokines. Fibrinogen is converted to fibrin by platelet factors, platelet-stored molecules, and procoagulants. The clot is composed of fibrin, fibronectin, platelets and trapped blood cells, and facilitates the movement of other cellular components into the wound.

## (2) Inflammation

At the injury site, neutrophils marginate on the vessel wall, and migrated to the injury. Monocytes and macrophages peak at 3 days after injury and begin to decrease by the fifth day. These are derived from the surrounding tissues and from the hematogenous monocytes, and they promote wound debridement and release factors that stimulate fibroblast migration and proliferation.

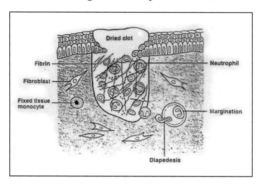

**Fig. 3.** Inflammation

### (3) Fibroplasias

Fibroblasts originate from mesenchymal cells of the blood vessel, and their movement is controlled by fibronectin, serotonin, and prostaglandins. Their movement peaks 2 days after injury, thereby preventing infection of the tissue. They are attracted to the wound by fibronectin and tissue hormones such as serotonin and prostaglandins. They appear on the 3rd day and become the dominant cellular component by the 5th day. They produce collagen, elastin, and mucopolysaccharides. Angiogenesis is immediately followed the movement of fibroblasts into the wound. Fibrovascular tissue is formed as consequence of fibroblast proliferation and angiogenesis.

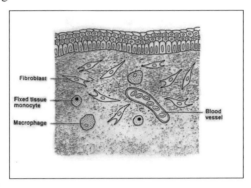

**Fig. 4.** Fibroplasia

### (4) Matrix remodelling

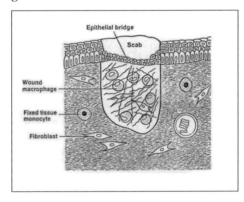

**Fig. 5.** Matrix remodeling

Myofibroblasts originate from fibroblasts, synthesize collagen at 5 to 7 days after injury and peak at 4 to 5 weeks after wounding. Centripetal movement of the wound edge is mediated by myofibroblasts. Dehydration and condensation begins during the fibroblastic phase and may last for more than a year. Collagen matures and is transformed into type I. The number of fibroblasts and blood vessels decreases, and a small, dense scar is formed.

**Table 1.**    Various cytokines involved in wound healing (Choi and Chung 2003)

| Cytokine | Symbol | Source | Functions |
|---|---|---|---|
| Platelet-derived growth factor | PDGF | Platelets, macrophages, endothelial cells, smooth muscle cells | Fibroblast proliferation, chemotaxis, collagen metabolism: chemotaxis and activation of neutrophils, macrophages: angiogenesis |
| Transforming growth factor β | TGF-β | Platelets, neutrophils, lymphocytes, macrophages, and many other tissues and cells | Fibroblast proliferation, chemotaxis, collagen metabolism, chemotaxis, indirect angiogenesis, and action of other growth factors |
| Epidermal growth factor | EGF | Platelets, saliva, urine, milk, plasma | Stimulate epithelial cell and fibroblast proliferation and granulation tissue formation |
| Transforming growth factor | TGF-α | Activated macrophages, platelets, keratinocytes, and many tissues | Similar to EGF functions |
| Interleukins | IL-1 | Macrophages, lymphocytes, and many other tissues and cells | Fibroblast proliferation, collagenase, neutrophil chemotaxis |
| Tumor necrosis factor | TNF | Macrophages, mast cells, T lymphocytes | Fibroblast proliferation |
| Fibroblast growth factor | FGF | Brain, pituitary, macrophages, and many other tissues and cells | Fibroblast and epithelial cell proliferation : stimulates matrix deposition, wound contraction and angiogenesis |
| Keratinocyte growth factor | KGF | Fibroblasts | Epithelial cell proliferation |
| Insulin-like growth factor-I | IGF-1 | Liver, plasma, Fibroblasts | Stimulate synthesis of sulfated proteoglycan, collagen, and fibroblast proliferation |
| Human growth hormone | huGH | Pituitary, plasma | Anabolism, stimulates IGF-1 |
| Interferons | IFN | Lymphocytes, Fibroblasts | Inhibition of fibroblast proliferation and collagen synthesis |

### 4.1.2. CAP project on the wound healing effect of Aloe vera components

Aloe has long been known for its various pharmacological effects. However, because the relationship between the aloe components and these effects has not been well elucidated, this study was designed to single out the most effective component, to examine its effect in vitro, and to confirm its effect in *in vivo* experimental animal models.(Choi *et al.* 2001) A glycoprotein fraction of 5·5 kd was isolated from aloe vera. It was identified as a glycoprotein by SDS-PAGE and periodic acid - Schiff staining, TLC using carbohydrate specific plates, spectroscopy, HPLC, and amino acid analysis. It exhibited a significant cell proliferation-stimulatory effect. It also accelerated the recovery of an artificial wound on a monolayer of normal keratinocytes, enhanced the thickening of the epidermal covering with an overall appearance of proliferating phenotype, and increased the expression of EGF receptor, fibronectin, fibronectin receptor, keratin 5/14, and keratin 1/10. The cell proliferation-stimulatory activity was further confirmed by enhanced wound healing in hairless mice.

### (1). Materials and Methods
### Preparation of aloe fractions

*Aloe vera* was fractionated by Dr. Byung-Wha Son (Bukyung National University, Busan Korea), and the glycoprotein fraction, G1G1M1DI2, was used in this experiment.

### Thymidine uptake assay

Fractions isolated from *Aloe vera* were evaluated for cell proliferating activity by measuring the thymidine uptake of human squamous cell carcinoma 13 (SCC13) cells in the presence of each aloe fraction. SCC13 cells were seeded in a 96-well tissue culture plate at 5,000 cells per well, and fed a media of Dulbecco's modified Eagle's medium (Gibco BRL, Grand island, NY, USA) and Ham's nutrient (Gibco BRL) mixture (3 : 1 by weight), supplemented with 10% fetal bovine serum (FBS; Hyclone, Logan, UT, U.S.A.). When the cell population reached 80% of confluency, an aloe fraction was applied at 1 mg/ml with serum-free media and incubated for 24 h. Then [$^3$H]thymidine (Amersham, Arlington Heights, IL, U.S.A.) was incubated at 5 µCi/ml for 3 h. Unreacted thymidine was washed out by PBS. The nucleic acids were dried at room temperature for 15 min, and fixed in 10% trichloroacetic acid (TCA) at 4°C for 1 h. TCA was removed by washing with PBS, dried at room temperature for 15 min, solubilized in 0.5M NaOH for 6 h, neutralized with 0.5M HCl, collected and mixed with a scintillation cocktail (Packard, Downers Grove, U.S.A.), and the amount of thymidine uptake was measured by a scintillation counter (Packard).

### Preparation of keratinocytes

Human foreskins were obtained from newborn babies circumcised in newborn nurseries. Foreskins were washed extensively with multiple changes of PBS, subcutaneous tissue was removed, and the remaining samples were enzymatically dissociated in multiple changes of 0.25% trypsin and versene (50 : 50). Epidermal sheets were peeled from the dermis, minced, and dispersed in trypsin solution by repeating pipetting. Cell suspensions were pelleted from the trypsin solution, sequentially resuspended, and washed with PBS, by centrifugation at 1,000 g for 5 min at 20°C. These cells were raised in a tissue culture dish with mouse fibroblasts J2 as feeder cells (donated by Dr. E. Fuchs of University of Chicago, U.S.A.). Cultures were maintained with a growth medium consisting of Dulbecco's modified Eagle's medium and Ham's nutrient mixture F12 at a 3 : 1 ratio. The medium was supplemented with 10% FBS, 1 x $10^{-10}$M cholera toxin, 0·4 µg/ml hydrocortisone, 5 µg/ml insulin, 5 µg/ml transferrin, and 2 x $10^{-11}$M triiodothyronine. Cultures were fed every 3 days and subcultured by dispersal in 0.025% trypsin in PBS and replated at a split ratio of 1 : 3. Cultures were used between passages 2 and 3.

### Effect of G1G1M1DI2 on keratinocyte migration

Normal human keratinocytes were cultured and reached confluence in a 6-well culture plate, after which the culture medium was drained away. A wound (width 0.5 mm) was created on an area of cells by gentle scraping with a rubber stick and moving back and forth against the top of the culture. The wells were washed four times with PBS to remove any remaining cellular debris. Cultures were maintained with a medium supplemented with 5% FBS. The aloe fraction G1G1M1DI2 was added to the culture at 100 µg/ml, and the control culture received PBS. The wound restoration was photographed at 16, 25, and 40 h post injury.

### Effect of G1G1M1DI2 on epidermis formation from keratinocytes in raft-culture

When the cells reached 90% of confluency, they were trypsinized on a collagen matrix for raft-culture. Briefly, mouse fibroblasts obtained from Dr. Elaine Fuchs were mixed with Type-I collagen matrix (Cell Matrix, Nitta gelatin, Tokyo, Japan) at a density of 17,000 cells per Millericell, and seeded in 12 mm Millericell (Falcon, Lincoln Park, NJ, U.S.A.). Keratinocytes were then seeded on the matrix at 17,000 cells per Millericell. Cells were cultured submerged in media for 7 days, transferred to the air-liquid interface and then raised for 21 days. An aloe fraction of G1G1M1DI2 was applied to cells with serum-free media with 0, 0.05, 0.5, 50 µg /ml every two days. Part of the epidermal tissue formed by 3 weeks of culture was taken, fixed in Carnoy solution (ethanol: glacial acetic acid: chloroform = 6: 1: 3 by volume) and washed with 60% and 80% ethanol and put in paraffin blocks for morphological comparison.

### Immunohistochemical study

Slices (5 µm) of paraffin blocks were deparaffinized and hydrated before immunohistochemical staining. Immunostaining was carried out as described by Choi and Fuchs.(Choi Y and Fuchs E 1990) Antisera and dilutions were as follows: mouse monoclonal antisera against human EGF receptor, 1 : 5 (Triton diagnostics, Alameda, CA, U.S.A.); rabbit polyclonal antisera against human fibronectin receptor, 1 : 200 (Chemicon, Temecula, CA, U.S.A.); rabbit polyclonal antisera against human fibronectin, 1 : 100 (Biomedical Tech, Stoughton, MA, U.S.A.); rabbit poly-clonal antisera against human keratin 5/14, 1 : 100 (Chemicon); and mouse mono-clonal antisera against keratin 1/10, 1 : 50 (Chemicon). After incubation with primary antisera, the sections were then subjected to immunogold enhancement (Amersham). The immunostaining intensity was estimated by an Image Analyzer (BAS-2500, Fujifilm, Tokyo, Japan) with image analysis software (MCID ver 3.0, Imaging Research Inc., Ontario, Canada) and expressed as a percentage of the control.

### Effect of G1G1M1DI2 on wound healing in hairless mice

Ten male hairless mice (Crl:SKH1-hrBR, Charles River Lab., Wilmington, MA, U.S.A.) were used for each control and experimental group in this study. Under pentobarbital sodium (1.92 mg/kg) anesthesia, epithelial wounds (approximately 154 mm$^2$) down to subcutaneous fat were made with a biopsy punch on both sides of the gluteal region, approximately 1 cm distal from the vertebral column. Experimental animals were given G1G1M1DI2 (10 mg/g ointment) as ointment with gentamicin 0.1% everyday. The control animals received the cream only. The ointment was changed daily to permit evaluation of the size and appearance of each wound until the wound became completely healed. The wound area was photographed, and the size of the wound was estimated using an Image Analyzer (BAS-2500, Fujifilm) with image analysis software (MCID ver 3.0, Imaging Research Inc.) and expressed as a percentage of wound size at day 0 post injury. On day 8, a 7-mm diameter punch of the skin was immersed in cold, neutral, buffered formalin for 24 h, embedded in paraffin, sectioned into 5 μm slices, and stained using hematoxylin and eosin.

## (2) Results

### Characterization of proliferation stimulating    glycoprotein fraction G1G1M1DI2

The G1G1 fraction obtained from the gel (G1) of *Aloe vera* was applied to an Amberlite XAD-2 column. The elute (G1G1M1) with 50% aqueous methanol showed activity in the thymidine uptake assay. It was subjected to ion-exchange column separation and afforded a non-binding fraction (G1G1M1DI1) and binding fraction (G1G1M1DI2) (Fig. 6). G1G1M1DI1 did not show any activity in the thymidine uptake assay, but the main activity was found in the binding fraction G1G1M1DI2 (Fig. 6). The amount of [$^3$H] thymidine taken up by SCC13 cells was measured to screen and pick out the most effective fraction. Increased activity was observed from the highly purified aloe fractions. Aloe components prepared from fraction G1 showed steady thymidine uptake which increased with further purification while G1G1M1DI2 showed an approximate 6.5-fold increase in thymidine uptake compared to the G1 fraction.

### Effect of G1G1M1DI2 on migration of keratinocytes

Because cell migration is involved during wound healing, we investigated whether G1G1M1DI2 enhances cell locomotion by evaluating cell migration in the presence of an aloe fraction (Fig. 7). Prior to wounding, the cells appeared generally flat, polygonal in shape, and relatively regular in size. Immediately after wounding, there were no cells within the defective area. G1G1M1DI2 was applied at 100 µg/ml after which cleavage recovery was observed. At 16, 25, and 40 h after incision, cells in the G1G1M1DI2-treated group had multiplied and become more compact. Membranes of adjacent cells became tightly juxtaposed, indicating close cell-to-cell contact. Under this condition, the leading edge of the closing wound was proceeded by a number of migrating elongated cells. Cells more distal to the leading edge appeared confluent and generally polygonal in shape. There appeared to be a larger population of migrating cells throughout the original wound area than in the controls. After 40 h the cell migration finally filled the cleaved area. However, in the control group, cells were quite slow to rearrange and form permanent cell-cell contacts and did not restore the cleaved area. This result shows that G1G1M1DI2 exerted its effect in the cleavage recovery by attracting cells to the cleaved site as well as by stimulating the division of the cells. The migration assay was repeated in duplicate, in two independent experiments In both cases the G1G1M1DI2 treatment was found to result in the accelerated migration of keratinocytes. Next, G1G1M1DI2 was evaluated for its capability to augment epidermal tissue formation from keratinocytes and to increase the expression of cell proliferation-related factors.

**Effect of G1G1M1DI2 on the architecture of epidermis in raft-culture**

The stained paraffin sections of the artificial epidermis raised in the raft-culture were compared by staining with hematoxylin and eosin (Fig. 8). In the control most basal cells died and only a few cells survived. A very thin spinous layer and a granular layer were observed with stratum corneum. As the G1G1M1DI2 concentration was increased from 0.05 to 50 µg/ml, basal cells multiplied and formed thick epithelial coverings. At 0.05 µg/ml, three to four spinous and granular layers were observed. The number of cells containing a nucleus increased as the G1G1M1DI2 concentration was increased.   From 0.5 up to 50 µg/ml, the number of cells with a nucleus further increased, and a thicker stratum corneum was formed.

**Immunohistochemical comparison**

The effect of G1G1M1DI2 on the expression of protein markers related with cell proliferation was studied using immunohistochemical methods (Fig 9). Immunostaining against an EGF receptor in the control showed that a very low level of EGF receptor expression was observed in the basal layer. As the G1G1M1DI2 concentration increased, the receptor expression increased on the cell membrane of the basal layer and even in the innermost spinous cells. Consistent with basal cell increase shown by hematoxylin and eosin staining from a G1G1M1DI2 concentration of 0.5 to 50 µg/ml, EGF receptor expression was increased in a dose dependent manner to 113% and 220% ($\pm$ 8%) compared to the controls at a G1G1M1DI2 concentration of 0.5 and 50 µg/ml, respectively. In normal skin, fibronectin is essentially located at the level of the papillary dermis. In our study, mouse fibroblasts embedded in the collagen matrix produced fibronectin, which made it possible for cells to locomote through the interaction with fibronectin receptors. Fibronectin was expressed in a dose-dependent manner. Its expression was increased to 294% (± 34%) and 408% (± 80%) compared to the controls at a G1G1M1DI2 concentration of 0.5 and 50 μg/ml, respectively. The expression of fibronectin receptor was also increased in a dose-dependent fashion to 159% (± 11%) and 220% (± 19%) compared to the controls at a G1G1M1DI2 concentration of 0.5 and 50 μg m/L, respectively. Cells expressing the fibronectin receptor were located at the basal and suprabasal layers of the artificial tissue. Fibronectin receptor was also found between the collagen matrix and the basal layer. Expression of keratins was examined in the artificial tissue. Keratin 5/14 expression in the basal and spinous layers was increased in a dose dependent manner to 260% (± 11%) and 308% (± 17%) compared to the controls at a G1G1M1DI2 concentration of 0.5 and 50 μg/ml, respectively'). However, keratin 1/10 was expressed in suprabasal and granular cells, and its expression increased to 228% (± 24%) and 174% (± 19%) compared to the controls at a G1G1M1DI2 concentration of 0.5 and 50 μg/ml, respectively.

**Effect of G1G1M1DI2 on wound healing in hairless mice**

The cell-proliferating activity of G1G1M1DI2 was investigated by examining whether or not it enhanced wound healing in hairless mice. The result revealed that daily application of the aloe glycoprotein fraction induced an increase in new epithelial area and that the treated lesion healed faster than the controls (Fig. 10). Healing was accompanied by enhanced granulation and increased epithelization, starting from the periphery of the wound. At day 4, 6, and 8 the wound area was recovered by 5.9, 84.2, and 98.9% of the original size in the G1G1M1DI2 treated group, compared to by 4.4, 49.5, and 69.5%, in the controls, respectively. Histological sections of the recovered area at day 8 showed that in the controls one basal layer was observed with one to two layers of spinous and granular layers. However, in the G1G1M1DI2 treated mice three layers of the basal layer were observed with an increased number of nucleus-containing cells, and more spinous and granular layers.

**(3) Discussion**

Regarding the cell proliferation activity by whole extract or components from *Aloe vera*, aloesin, mannose-6-phosphate, glycoprotein, and aloe-emodin have been reported to have such an activity. Recently, Yagi *et al.*(Yagi *et al.* 1997) also reported that a glycoprotein fraction of 29 kd isolated from the *Aloe barbadensis* Miller has proliferation-promoting activity. It is composed of two subunits each of 14 kd. However, smaller proteins were neither observed nor searched for during this study. At present, it is not possible to compare the homologies of the 29 kd protein and our 5·5 kd protein, due to the lack of published information on the 29 kd protein. It is possible that *Aloe vera* contains many cell proliferation-stimulatory components with varying activities. In our study G1G1M1DI2 was probably not the only active component of *Aloe vera*, because other fractions also exhibited cell proliferation activity.

Various factors are known to affect cell migration including EGF,(Nilsson *et al.* 1995) IL-1β,(Cumberbatch *et al.* 1997) nitric oxide, (Noiri *et al.* 1996) platelet-derived growth factor,(Nelson *et al.* 1997) and short-chain fatty acids. (Wilson and Gibson 1997) TGF-β has controversial effects as it either enhances or inhibits the migration. (Grant *et al.* 1992; Ilio *et al.* 1995) Maehiro *et al.*(Maehiro *et al.* 1997) reported that epithelial restoration by EGF is done by stimulating both migration and proliferation, while Nilsson *et al.* (Nilsson *et al.* 1995) showed that growth factor-induced migration of thy-rocytes is not strictly coupled to the mitogenic activity. It is likely that the growth factors stimulate cell migration depending on the cell type and tissue. In our study the glycoprotein fraction G1G1M1DI2 accelerated cell migration. It is likely that cell migration was concomitant with cell proliferation which was enhanced by the aloe fraction.

There were a few cells that survived in the absence of serum and aloe fraction G1G1M1DI2 in the raft-culture. However, in the presence of G1G1M1DI2, keratinocytes survived and multiplied, even in the serum-free media, and formed an epidermis with an overall appearance of the proliferating phenotype. This indicates that this aloe fraction G1G1M1DI2 has a cell proliferation stimulatory effect.

EGF receptor mediates paracrine and autocrine growth regulation of normal and malignant cells. (Cook *et al.* 1993) Stoscheck *et al.* (Stoscheck *et al.* 1992) demonstrated that an increase in the number of EGF receptors precedes the hypertrophic response. The aloe fraction G1G1M1DI2 enhanced the DNA syn-thesis as well as the EGF receptor expression. This suggests that EGF receptors transmitted the cell proliferation signal from G1G1M1DI2. Another possibility is that G1G1M1DI2 may activate the general metabolism, thereby enhancing meta-bolic activities with a concomitant increase of EGF receptor.

There was a significant increase of EGF receptor expression in the artificial epidermis raised in raft-culture. It was suggested that the overexpression of EGF receptors may be in part responsible for the tumorigenic potential of cutaneous epithelial malignancies. (Groves *et al.* 1992) Furthermore, it was reported that there were twice as many EGF receptors in the squamous cell carcinoma of the head, neck, lung, cervix and skin as in the other organ. (Ozanne *et al.* 1986) Contrary to this, Tomatis *et al.* suggested that the cell-proliferation effect is not necessarily correlated with metastasis in keratinocytes proliferation. (Tomatis 1993) Jacobberger *et al.* (Jacobberger *et al.* 1995) showed that the pro-liferation rate was decreased by TGF-β1 in cervical epithelial cells, while EGF receptor levels were increased. Ottensmeier *et al.* (Ottensmeier *et al.* 1996) suggested that EGF receptor activation through autocrine pathways is not a major mechanism for the growth of many ovarian cancer cell lines. In our study, although the EGF receptor was significantly increased by G1G1M1DI2 in the artificial tissue raised in the raft-culture, keratinocytes did not invade the matrix, but entered the terminal differentiation as *in vivo* when the stratified epidermal cells were exposed to air, as suggested by Tomatis *et al.* (Tomatis 1993) G1G1M1DI2 did not exhibit uncontrolled cell proliferation.

Human keratinocytes express several receptors of the integrin family. Expression is normally confined to the basal layer of keratinocytes, both in mature epidermis and during development. (Watt *et al.* 1993) Ohashi *et al.* (Ohashi *et al.*

1995) reported that increased attachment of corneal epithelial cells to fibro-nectin and enhanced migration on corneal stroma by IL-6 are due to temporal up-regulation of $\alpha5\beta1$ expression in corneal epithelial cells. Nista *et al*. (Nista *et al*. 1997) reported that fibronectin delivers a mitogenic signal to human mammary carcinoma cells. The enhanced expression of fibronectin receptor is also observed in wound healing and cellular activation. (Ra *et al*. 1994) In our study, the fibro-blasts seeded in the collagen matrix of the raft-culture produced fibronectin in proportion to the glycoprotein G1G1M1DI2 concentration. In addition, the kerati-nocytes in the basal layer of the raft-culture expressed fibronectin receptor in a dose dependent manner. This confirms that G1G1M1DI2 stimulated both the fibroblasts and the keratinocytes in producing fibronectin and its receptor.

A keratin pair represents the most sensitive marker for the epidermis state. (Nelson and Sun 1983; Dale *et al*. 1978) Significant changes were observed in the raft-culture as the expression of keratin 5/14 was enhanced proportionally with respect to the G1G1M1DI2 concentration, whereas the synthesis of keratin 1/10 was slightly decreased at higher G1G1M1DI2 concentrations. This indicates that G1G1M1DI2 affects cell proliferation rather than differentiation.

Moreover, in the control no immunostaining of keratin 1/10 was observed, which might be due to undetectable levels of expression of the keratin in the sur-viving cells or the death of most basal cells in the absence of either serum or glycoprotein fraction G1G1M1DI2. Keratin 6/16 is a marker of regeneration-associated differentiation and is induced in epidermal diseases associated with hy-perproliferation. (Smedts *et al*. 1993) However, the expression of keratin 6/16 was not examined in this study, because we studied the effect of G1G1M1DI2 in artifi-cial epidermis raised in raft-culture, not in regenerated tissue or in the diseased epidermal tissue associated with hyperproliferation. Furthermore, it was our inten-tion to observe the dose dependent cell proliferation effect of G1G1M1DI2. Therefore, the dose dependent expression of keratin 5/14 was studied in compari-son to that of keratin 1/10. However, we do not exclude the possibility that keratin 6/16 is increased in artificial tissue, because the condition in raft culture bears some similarity with that in wounds. (Wilson AJ and Gibson 1997).

The hairless mouse was selected as a wound model for epidermal injury because it offers two advantages. First, the epidermis does not have a fur coat, which inter-feres with the separation of epidermis from dermis. Second, the size and economy of a hairless mouse makes it an ideal model to evaluate the effects of pharma-cologic agents on the wound-healing process. In the wounded hairless mouse, G1G1M1DI2 treatment stimulated keratinocyte proliferation and thus shortened the healing time, which was demonstrated by several basal and suprabasal layers with well proliferating cells.

For the wound healing experiment, G1G1M1DI2 was used. Other fractions (G1, G1G1, G1G1M1) were not examined, because we were apprehensive that any in-hibitory components possibly residing in the other crude fractions may decrease the wound healing effect of G1G1M1DI2. (Yagi *et al*. 1997) It is possible that the conflicting effects of G1G1M1DI2 and of the other inhibitory components in-duced variability in the results of the therapeutic experiments.

Although controversy remains regarding the wound healing effect of whole extracts of aloe, this report clearly demonstrates that the glycoprotein fraction G1G1M1DI2 enhances keratinocyte multiplication and migration, proliferation-related factor expression, and epidermis formation, thereby leading to wound healing.

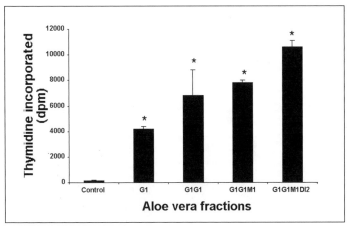

**Fig. 6.** Characterization of glycoprotein fraction G1G1M1DI2. Effect of various fractions of *Aloe vera* on [³H] thymidine uptake by squamous cell carcinoma 13 cells. Data represent mean ±SD of three experiments. *P<0.05 compared with control, in which an aloe fraction was not included in the thymidine uptake assay (Student's t-test). G1, G1G1, G1G1M1 and G1H1M1DI2 are described in the Materials and Methods. dpm, disintegrations per min.

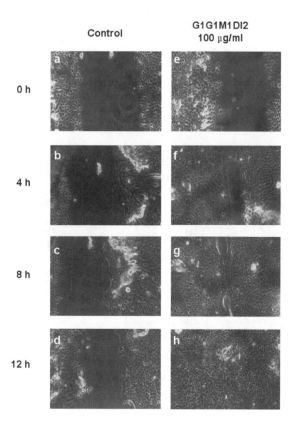

**Fig. 7.** Effect of glycoprotein fraction G1G1M1DI2(100ug/ml) on the migration of human keratinocytes on a monolayer. (a-d) Control: (e-h) G1G1M1DI2-treated group. This experiment was repeated in duplicate in two independent experiments. Photographs were taken at 0, 16, 25 and 40 h after injury (original magnification ×100)

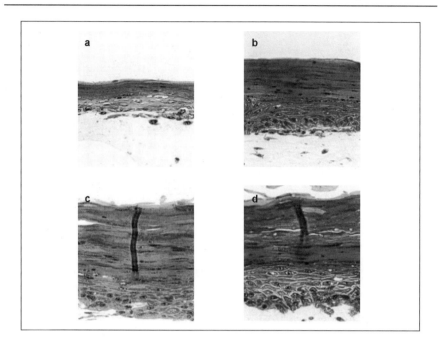

**Fig. 8**. Effect of glycoprotein fraction G1G1M1DI2 on the formation of epidermis from keratinocytes in raft culture. Cultures were treated with G1G1M1DI2 for 21 days immediately after exposure to the air. Cultures were grown (a) in the absence of G1G1M1DI2, or in the presence of G1G1M1DI2 at a concentration of (b) 0.05, (c) 0.5 or (d) 50 ug/ml (haematoxylin and eosin, original magnification×200). (See Plate 5.)

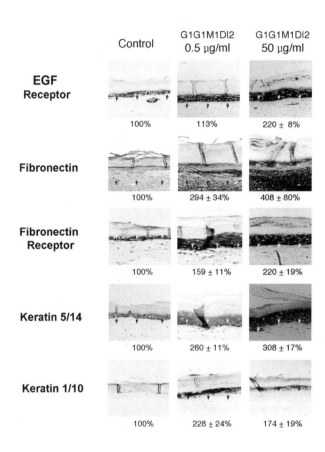

**Fig. 9**. Immunohistochemistry of the artificial epidermis raised in the raft culture. Immunohistochemical staining was used to detect epidermal growth factor (EGF) receptor, fibronectin, fibronectin receptor, keratin 5/14 and keratin 1/10 in the artificial epidermis raised in the absence or presence of 0.5 to 50 ug/ml G1H1M1DI2. Immunostaining is indicated by arrowheads. Data below each picture represent mean ± SD of relative percentage of optical density of three experiments. The immunostaining intensity was estimated by an image analyser with image analysis software (immunogold staining, original magnification × 200). (See Plate 6.)

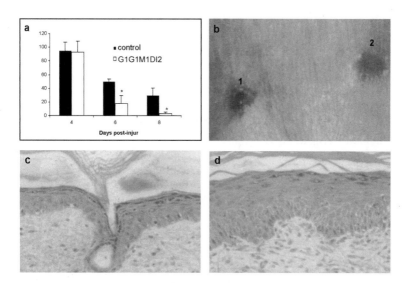

**Fig. 10.** Effect of glycoprotein fraction G1G1M1DI2 on the wound healing of hairless mice. (a) Data represent relative percentage of wound size (mean±SD; n=10). The wound area was photographed and the  size of the wound was estimated using a BAS-2500 image analyser with image analysis software MCID ver. 3.0 and expressed as a percentage of wound size at day 0 after injury. *$P<0.05$ compared with control (Student's t-test). (b) Wound area on day 10 after treatment: 1, control; 2, G1G1M1DI2-treated group (10 mg/g ointment). (c, d) Morphology of the wound area at day 10 after treatment: (c) control (d) G1G1MIDI2-treated group (haematoxylin and eosin, original magnification×200). (See Plate 7.)

## References

1. Burkitt HG, Young B, Heath JW (1993) Functional Histology Churchill Livingstone 153-169.
2. Choi S, Chung MH (2003) A review on the relationship between *Aloe vera* components and their biologic effects Seminars in Integrative Medicine 1(1) 53-62.
3. Choi SW, Son BW, Son YS, Park YI, Lee SK, Chung MH (2001) The wound healing effect of a glycoprotein fraction isolated from *Aloe vera* British Journal of Dermatology 145:535-545.
4. Choi Y, Fuchs E (1990) TGF-beta and retinoic acid regulators of growth and modifiers of differentiation in human epidermal cells *Cell Regul* 1 791-809.
5. Cumberbatch M, Dearman RJ, Kimber I (1997) Interleukin 1 beta and the stimulation of Langerhans cell migration: comparisons with tumor necrosis factor alpha *Arch Dermatol Res* 289: 277-84.
6. Dale BA, Holbrook KA, Steinert PM (1978) Assembly of stratum corneum basic protein and keratin filaments in macrofibrils *Nature* 27: 729-31.
7. Grant MB, Khaw PT, Schultz GS, Adams JL, Shimizu RW (1992) Effects of epidermal growth factor, fibroblast growth factor, and transforming growth factor-beta on corneal cell chemotaxis *Invest Ophthalmol Vis Sci* 33 3292-301.
8. Groves RW, Allen MH, MacDonald DM (1992)Abnormal expression of epidermal growth factor receptor in cutaneous epithelial tumours *J Cutan Pathol* 19: 66-72.
9. Ilio KY, Sensibar JA, Lee C (1995) Effect of TGF-beta 1, TGF-alpha, and EGF on cell proliferation and cell death in rat ventral prostatic epithelial cells in culture *J Androl* 16: 482-90.
10. Jacobberger JW, Sizemore N, Gorodeski G, Rorke EA (1995) Transforming growth factor beta regulation of epidermal growth factor receptor in ectocervical epithelial cells *Exp Cell Res* 220: 390-6.
11. Maehiro K, Watanabe S, Hirose M, Iwazaki R, Miwa H, Sato N (1997) Effects of epidermal growth factor and insulin on migration and proliferation of primary cultured rabbit gastric epithelial cells *J Gasteroenterol* 32: 573-8.
12. Nelson PR, Yamamura S, Kent KC (1997) Platelet-derived growth factor and extracellular matrix proteins provide a synergistic stimulus for human vascular smooth muscle cell migration. *J Vasc Surg* 26: 104-12.
13. Nelson WG, Sun TT(1983) The 50- and 58-kdalton keratin classes as molecular markers for stratified squamous epithelia: cell culture studies. *J Cell Biol* 97: 244-51.
14. Nilsson M, Dahlman T, Westermark B, Westermark K (1995) Transforming growth factor-beta promotes epidermal growth factor-induced thyroid cell migration and follicle neoformation in collagen gel separable from cell proliferation. *Exp Cell Res* 220: 257-65.
15. Nista A, Leonetti C, Bernardini G, Mattioni M, Santoni A (1997) Functional role of alpha4beta1 and alpha5beta1 integrin fibronectin receptors expressed on adriamycin-resistant MCF-7 human mammary carcinoma cells. *Int J Cancer* 72: 133-41.

16. Noiri E, Peresleni T, Srivastava N, Weber P, Bahou WF, Peunova N, Goligorsky MS (1996) Nitric oxide is necessary for a switch from stationary to locomoting phenotype in epithelial cells. *Am J Physiol* 270 (3Pt1): C794-802.
17. Ohashi H, Maeda T, Mishima H, Otori T, Nishida T, Sekiguchi K (1995) Up-regulation of integrin alpha 5 beta 1 expression by interleukin-6 in rabbit corneal epithelial cells. *Exp Cell Res* 218: 418-23.
18. Ottensmeier C, Swanson L, Strobel T, Druker B, Niloff J, Cannistra SA (1996) Absence of constitutive EGF receptor activation in ovarian cancer cell lines. *Br J Cancer* 74: 446-52.
19. Ozanne B, Richards CS, Hendler F, Burns D, Gusterson B (1986) Over-expression of the EGF receptor is a hallmark of squamous cell carcinomas. *J Pathol* 149: 9-14.
20. Pittelkow MR, Cook PW, Shipley GD, Derynck R, Coffey RJ Jr (1993) Autonomous growth of human keratinocytes requires epidermal growth factor receptor occupancy. *Cell Growth Differ* 4: 513-21.
21. Ra C, Yasuda M, Yagita H, Okumura K (1994) Fibronectin receptor integrins are involved in mast cell activation. *J Allergy Clin Immunol* 94: 625-8.
22. Smedts F, Ramaekers F, Leube RE, Keijser K, Link M, Vooijs P (1993) Expression of keratins 1, 6, 15, 16, and 20 in normal cervical epithelium, squamous metaplasia, cervical intraepithelial neoplasia, and cervical carcinoma. *Am J Pathol* 142: 403-12.
23. Stoscheck CM, Nanney LB, King, LE Jr (1992) Quantitative determination of EGF-R during epidermal wound healing. *J Invest Dermatol* 99: 645-9.
24. Tomatis L (1993) Cell proliferation and carcinogenesis: a brief history and current view based on an IARC workshop report *Environ Health Perspect* 101 (suppl 5) 149-51.
25. Watt FM, Jones PH (1993) Expression and function of the keratinocyte integrins *Dev* suppl 185-92.
26. Wilson AJ, Gibson PR (1997) Short-chain fatty acids promote the migration of colonic epithelial cells in vitro *Gastroenterology* 113: 487-96.
27. Yagi A, Egusa T, Arase M, Tanabe M, Tsuji H (1997) Isolation and characterization of the glycoprotein fraction with a proliferation-promoting activity on human and hamster cells in vitro from *Aloe vera* gel *Planta Med* 63 18-21.

## 4.2 Angiogenic effect

**Kim, Kyu Won, Ph.D.**

College of Pharmacy, Seoul National University

Aloe plants, a member of the lily family, are useful as food, and have occupied a prominent place in folk medicine for centuries. Aloes are herbaceous, shrubby, and xerophytic plants. They range in size from less than 30cm to over 15m in height and grow in a variety of terrains from sea level to as high as 2700m. The leaves are spiny, thick, and fleshy. The oldest recorded medicinal knowledge of aloe is its application to inflammatory states and use in catharsis in India. Aloe contains a number of anthraquinone glycosides, with the principal one being barbaloin (aloe-emodin anthrone C-10 glucoside) which has been used as a cathartic.

Only a few of more than 300 known aloe species are currently used by the pharmaceutical and cosmetic industries. Among them, *Aloe barbadensis*, commonly called *Aloe vera*, has been one of the most widely used healing plants in the history of mankind. *Aloe vera* is relatively unique compared to most other species of botanicals. In most cases that a botanical is identified as having properties associated with healing or cosmetic applications, it is typically a single ingredient or constituent which is isolated from the plant that is found to be responsible for the observed benefits. *Aloe vera*, on the other hand, is known to contain over 100 separate ingredients or constituents in the leaf and the mucilaginous gel inside the leaf. It has been suggested that the extract of *Aloe vera* promotes wound healing through the complex synergistic interaction of many substances including vitamin, amino acids, and other small constituent molecules.

*Aloe vera* gel is a clear, semi-solid, jelly-like material, which serves as the water storage organ for this desert plant. This jelly-like material is highly recommended for burns and stomach problems. The clinical evidence has repeatedly demonstrated that *Aloe vera* gel has significant ability in promoting vascularizing, and in reducing edema and inflammation. In addition, it was reported that *Aloe vera* gel improved wound healing in a dose-dependent fashion. However, the critical pharmacological mechanism which might justify the therapeutic use of *Aloe vera* gel as a healing agent is still unknown.

In wound healing, angiogenesis is an essential process. There are two processes implicated in the formation of new capillaries: vasculogenesis and angiogenesis. Vasculogenesis is the *de novo* differentiation of mesodermal precursor cells into endothelial cells and the subsequent formation of a primary capillary plexus. Angiogenesis is the growth of new capillaries from pre – existing capillaries and post – capillary venules. While vasculogenesis appears to be limited to the early embryo, angiogenesis is known to occur throughout life. In addition to its role during development, angiogenesis occurs in physiological settings such as corpus luteum formation, and during wound healing.

Wound healing is presented as three overlapping series of events: inflammation, new tissue formation and matrix remodeling. (Dunphy *et al.* 1974).

Angiogenesis is required to furnish the new tissue formed in wound healing process with oxygen and metabolites and to dispose of the waste products of metabolism (Thompson *et al.* 1991). Thus it could be supposed that the wound healing effect of *Aloe vera* gel may be related to its angiogenic activity.

Angiogenesis, a process of formation of new capillaries from pre-existing vasculature is necessary to many physiological and pathological conditions including embryonic development, wound healing, tissue regeneration and tumor growth. (Folkman *et al.* 1987) Among the reported angiogenic activators, some have an ability to promote wound healing. For example, transforming growth factor (TGF) has been shown to increase the rate of healing of linear incision wounds in animal models and also enhance angiogenesis. (Amento and Beck 1991; Lee *et al.* 1998; Moon *et al.* 1999) In addition, local application of basic fibroblast growth factor (bFGF) to skin wounds has been reported to enhance both dermal as well as epidermal wound healing. (McGee *et al.* 1998; Hebda *et al.* 1990; Tsuboi *et al.* 1990) When angiogenesis is impaired in the aged or in the irradiated tissues, wound healing is retarded or unsuccessful. (Phillips *et al.* 1991) Thus, angiogenesis might be a key regulating process in order to heal wounds.

A number of pharmaceutical publications have eulogized the ability of *Aloe vera* gel to promote the healing of burns and other cutaneous injuries and ulcers. (Klein and Penneys 1988; Lushbaugh *et al.* 1953) *Aloe vera* gel improved wound healing in a dose-dependent fashion, and reduced edema and pain. (Davis *et al.* 1987, 1989) Based on this historical and scientific background, the research on angiogenic activity of *Aloe vera* gel is a cogent investigation. Thus, the purpose of this study is to elucidate the pharmacological mechanism of the wound healing effect of *Aloe vera* gel. This study is focused on the relationship between wound healing and angiogenesis.

In this study, *Aloe vera* gel was extracted via the specific fractionation procedure (Fig.1). Crude extracts (G1) were prepared by lyophilization of *Aloe vera* gel. In order, methanol extracts (G1M1), dichloromethane extracts (G1M1D1), and 90% methanol extract (G1M1D1M1) were obtained from G1. G1 was extracted three times with hot methanol. Methanol extracts (G1M1) were obtained from the methanol layer through evaporation under reduced pressure to dryness. G1M1 was dissolved in water and extracted three times with dichloro-methane to yield the dichloromethane layer. The dichloromethane layer was evaporated to make dichloromethane extracts (G1M1D1) under reduced pressure. G1M1D1 was refractionated with hexane/90% methanol to yield the 90% methanol layer. The methanol extract (G1M1D1M1) was obtained from 90% methanol layer by vacuumed-evaporation and chromatographed over a silica gel column with dichloromethanol elution.

First of all, to assess the invasiveness of endothelial cells, *in vitro* invasion assay was performed. Angiogenesis is associated with dramatic changes in endothelial cell behavior. Endothelial cells are mesodermally derived simple squamous epithelial cells arising from splanchnic mesoderm. In response to angiogenic stimuli, the normally quiescent microvascular endothelial cells start to invade, and migrate through the surrounding extracellular matrix. Subsequently, the proliferation and differentiation of the endothelial cells into tube-like structures occurs. It is well known that a crucial event during angiogenesis is the invasion of the perivascular extracellular matrix by sprouting endothelial cells while the basement membrane creates a critical barrier to passage of cells. A variety of *in vitro* systems have been developed to assess the invasiveness of the cells. An *in vitro* system allows the rapid and quantitative assessment of invasiveness and a means to screen for drugs. The assay was carried out following the method of Saiki *et al.* (Saiki *et al.* 1993) As shown in Fig.2, G1M1D1M1 stimulated the invasion of calf pulmonary artery endothelial (CPAE) cells in dose-dependent manner and reconstituted the basement membrane. (Saiki *et al.* 1993)

To further investigate whether *Aloe vera* gel induces angiogenesis, chorioallantoic membrane (CAM) assay was performed (Fig.3). Many substances have been identified showing angiogenic activities in angiogenesis assay systems such as the rabbit cornea and chick embryo chorioallantoic membrane assays. Among them, the CAM assay is useful for obtaining large amounts of data as screening studies to test the angiogenic activity of a variety of compounds. CAM serves as the major vehicle of gas exchange and the repository for secretory wastes during embryonic development. The respiratory function is provided by the extensive capillary network of CAM. CAM has been extensively used as a host for transplantation of a variety of normal and neoplastic tissues. The ability of CAM to support the growth of such tissues apparently increases as the membrane ages, reaching an optimum from 9 to 12 days. Because of the distinctive arrangement of capillaries and other vessels in CAM, this tissue provides a good model for studying factors which influence vascular migration and pattern.

Thermanox coverslips were loaded with various doses of *Aloe vera* gel extracts or phorbol 12-myristate 13-acetate (PMA). After air-drying, loaded coverslips were then applied to the CAM surface of 9-day-old chick embryos. Three days later, an appropriate volume of 10% fat emulsion (10% Intralipose) was injected into the 12-day-old embryo chorioallantois and the neovasculature was observed under a microscope. CAM containing the hypervascular zone was counted as positive CAM. As summarized in Table 1, G1M1D1M1 was the most active fraction, demonstrating 90% activity at 200 µg/egg.

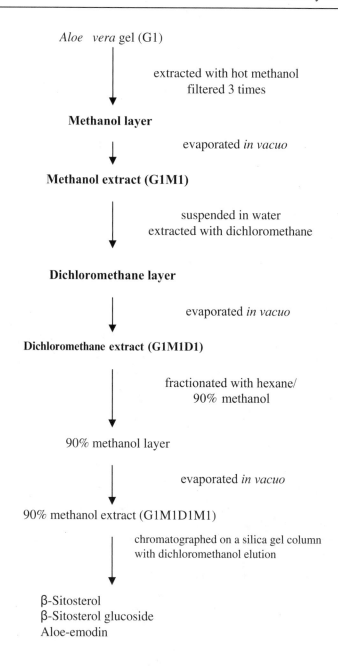

**Fig. 1.** Fractionation procedure of *Aloe vera* gel.

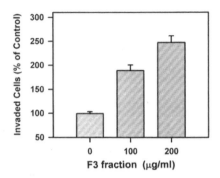

**Fig. 2**. Effects of G1M1D1M1 on the invasiveness of CPAE cells into Matrigel. F3 fractionmeans G1M1D1M1. CPAE cells were treated with desired doses of G1M1D1M1 for 48h. Invasion was determined by counting the number of the cells that had migrated to the lower side of the filter with optical microscopy.

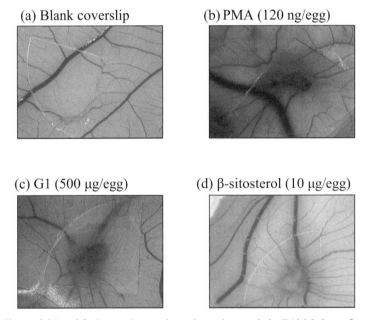

**Fig. 3**. Effects of G1 and β-sitosterol on embryonic angiogenesis in CAM 3 days after sample implantation. Fat emulsion (10%) was injected into chorioallantois to clear the vascular network. (a) Control CAMs treated with blank coverslips showed no angiogenic response. Positive responses were seen in CAMs implanted with coverslips loaded with (b) PMA (120 ng/egg), (c) G1 (500 μg/egg), and (d) β-sitosterol (10 μg/egg). (See Plate 8.)

Next, in order to identify the active compound having an angiogenic effect, G1M1D1M1 was purified into three compounds, β-sitosterol, β-sitosterol glucoside, and aloe-emodin, whose structures were identified by (Davis RH *et al.* 1989) C-NMR and [1]H-NMR. The chemical structures of these compounds are shown in Figure 4. The angiogenic activities of these compounds on CAM were also examined (Table 2). Induction of angiogenesis was observed in 64% of the eggs at 10 μg of β-sitosterol (Fig. 3d), but not in aloe-emodin or β-sitosterol glucoside. Moreover, co-application of aloe-emodin or β-sitosterol glucoside with β-sitosterol did not produce a higher degree of neovascularization as compared to β-sitosterol alone. As shown in Table 3, β-sitosterol stimulated angiogenesis in a dose-dependent manner.

To confirm the angiogenic capacity of β-sitosterol, the mouse Matrigel plug assay was performed. After injection into mice, the Matrigel plugs containing β-sitosterol alone were pale in color, indicating little or no neovascularization. Although heparin is not angiogenic itself, it has been reported to modulate angiogenic response. (Castellot *et al.* 1982; Taylor *et al.* 1982) Thus, the angio-genic activity of β-sitosterol in the presence of heparin was tested. Matrigel plugs supplemented with β-sitosterol and heparin were actually congested with blood.

Microscopic examination of β-sitosterol-containing Matrigel plugs stained with Masson Trichrome stain revealed the presence of abundant cellularity that was characteristically absent from control Matrigel plugs (Fig. 5Aa, b). The cells invading the β-sitosterol-containing plugs were organized to form tubular structures, often containing red cells (Fig. 5Ab). Measurement of hemoglobin content indicated the formation of a functional vasculature inside the Matrigel, which is the actual site of angiogenesis. Evaluation of the angiogenic response to β-sitosterol, in the presence of heparin, by measurement of hemoglobin content clearly showed an approximate increase compared with the vehicle control (Fig. 5B).

Based on the above experimental evidence for the angiogenic activity of β-sitosterol, a wound migration assay was performed to investigate the mechanism of the angiogenic process induced by β-sitosterol. Confluent cultures of human umbilical vein endothelial cells (HUVECs) were wounded with a razor blade. HUVECs were allowed to migrate for 24 h and were rinsed with serum-free medium. Photographs were taken through an inverted microscope (×40). The migration was measured as the number of cells moving into the wounded area after 24h (Fig. 6A). In Figure 6B, β-sitosterol stimulated the migration of HUVECs by nearly 40% compared with the control. In addition, the MTT assay was also done to study the effect of β-sitosterol on the proliferation of HUVECs. Compared with untreated cells, β-sitosterol did not promote the proliferation of endothelial cells. Taken together, these results indicate that β-sitosterol does not affect the proliferation, but stimulates the migration of HUVECs.

**Table 1.** Angiogenic activity of *Aloe vera* gel and its extracts

| Compounds | Dose(μg/egg) | Total No. tested In CAM assay | % positive |
|---|---|---|---|
| Control | - | 91 | 16 |
| PMA | 0.12 | 56 | 81 |
| G1 | 100 | 84 | 42 |
|  | 500 | 97 | 77 |
|  | 1000 | 83 | 94 |
| G1M1 | 250 | 19 | 42 |
|  | 500 | 13 | 92 |
| G1M1D1 | 50 | 36 | 62 |
|  | 250 | 40 | 83 |
|  | 500 | 43 | 90 |
| G1M1D1M1 | 50 | 45 | 75 |
|  | 100 | 35 | 80 |
|  | 200 | 50 | 90 |

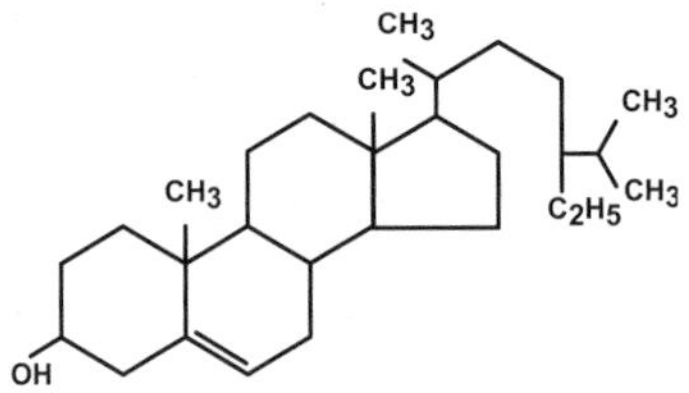

β –sitosterol 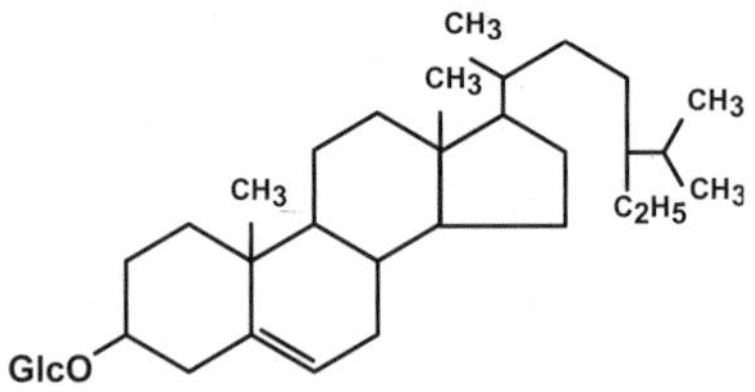                β -sitosterol glucoside

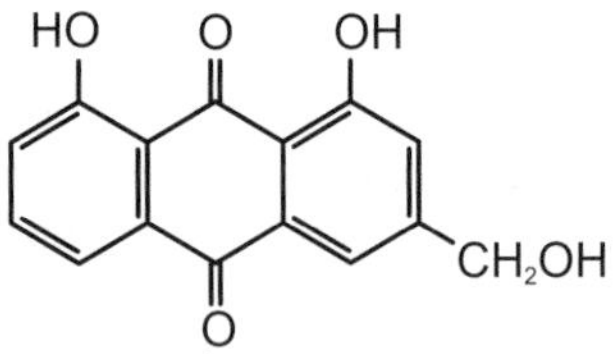

Aloe-emodin

**Fig. 4.** Structures of pure compounds from *Aloe vera*.

**Table 2.** Angiogenic activity of pure compounds isolated from *Aloe vera* gel

| Compounds | Dose (µg/egg) | Total No. tested In CAM assay | % positive |
|---|---|---|---|
| Control | - | 16 | 0 |
| PMA | 0.12 | 15 | 80 |
| aloe-emodin | 10 | 11 | 0 |
| β-sitosterol glucoside | 10 | 15 | 27 |
| β-sitosterol | 10 | 14 | 64 |
| β-sitosterol+ aloe-emodin | 10 | 12 | 50 |
| β-sitosterol + β-sitosterol glucoside | 10 | 14 | 64 |

**Table 3.** Angiogenic activity of β-sitosterol

| Compounds | Dose(µg/egg) | Total No. tested In CAM assay | % positive |
|---|---|---|---|
| Control | - | 30 | 0 |
| PMA | 0.12 | 30 | 82 |
| β-sitosterol | 5 | 27 | 56 |
| | 10 | 35 | 65 |
| | 20 | 32 | 76 |
| | 40 | 36 | 86 |

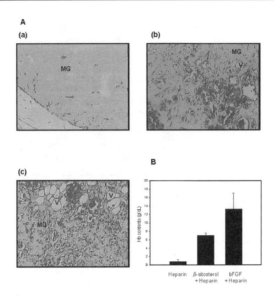

**Fig. 5.** A. Microscopic morphology of Matrigel plugs. Masson Trichrome-stained Matrigel plug impregnated with heparin (a); Masson Trichrome-stained Matrigel plug impregnated with β-sitosterol plus heparin (b); and Matrigel plug impregnated with bFGF plus heparin, as positive control (c). Male C57BL/6 mice were injected subcutaneously with either Matrigel containing heparin alone (40 unit/ml), heparin plus β-sitosterol (30 μg/ml) or heparin plus bFGF (100 ng/ml). Plugs were removed 5 days after injection and processed for histology. Magnification (100 X). B. Hemoglobin (Hb) content was measured by the Drabkin method using Drabkin reagent kit 525 (Sigma, St Louis, MO). The concentration of Hb was calculated from a known amount of Hb assayed in parallel. MG, Matrigel; v, vessels. (See Plate 9.)

*Aloe vera* gel has been known to promote wound healing (Lushbaugh *et al.* 1953). However, the mechanism of its healing activity is still unknown. Based on the fact that angiogenesis is a critical process for successful healing of wounds, it could be a reasonable hypothesis that *Aloe vera* contains compounds that stimulate angiogenesis. To test this hypothesis, it was investigated whether the crude extract of *Aloe vera* gel could induce angiogenesis on CAM after *in vitro* invasion assay for CPAE cells. The result that the crude extract of *Aloe vera* gel actively induced neovascularization on CAM provides support for the existence of some angiogenic compounds in *Aloe vera* gel. Therefore, the crude extract of *Aloe vera* gel was sequentially fractionated into methanol extract (G1M1), dichloromethanol extract (G1M1D1) and 90% methanol extract (G1M1D1M1). Thereafter, G1M1D1M1, the most active fraction, was purified and the structures of compounds were subsequently identified as aloe-emodin, β-sitosterol and β-sitosterol glucoside by [13]C-NMR and (Dunphy *et al.* 1974) H-NMR. The angiogenic activity of these fractions and the pure compounds were then compared by the CAM assay. Among them, β-sitosterol strongly induced angiogenesis compared to other compounds on

CAM, implicating that β-sitosterol is the major active component of *Aloe vera* gel which has the angiogenic activity.

β-sitosterol is one of the most common, plant-derived steroids with a structure similar to that of cholesterol. It was previously reported that β-sitosterol has diverse pharmacological effects. β-sitosterol inhibits the growth of HT-29 human colon cancer cells by activating the sphingomyelin cycle(Awad *et al.* 1998). A recent report suggested that β-sitosterol might have potential application in the drug therapy for severe hypercholesterolemia by lowering the level of plasma cholesterol, particularly of LDL cholesterol[19].Sitosterol has also been reported to enhance the production of plasminogen activator in cultured bovine carotid endothelial cells(Hagiwara *et al.* 1984; Shimonaka *et al.* 1984). Adding to the implication of these recent reports, a novel biological function of β-sitosterol for the stimulation of angiogenesis *in vivo* was demonstrated through this study.

**A**

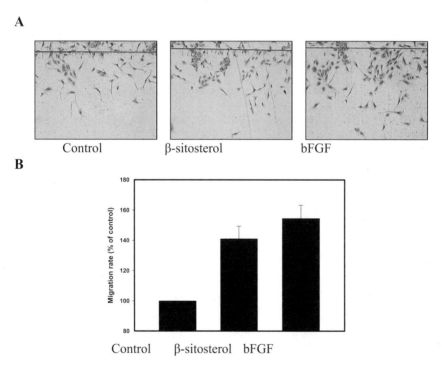

**B**

Fig. 6. Effects of β-sitosterol on the migration of HUVECs. HUVECs were seeded on a gelatin-coated culture dish. (A) At the 90% confluency, the monolayers were wounded with a razor blade. Wounded cells were rinsed with serum free medium and then incubated with β-sitosterol (2 μM) for 24h. bFGF (30 ng/ml) treated HUVECs were used as a positive control. The number of HUVECs that moved beyond the reference line was counted. (B) Results are expressed as percentages of control ± SEM. *, $P < 0.001$ compared with the control. (See Plate 10.)

In the CAM assay, β-sitosterol was able to induce angiogenesis by itself. However, β-sitosterol required heparin to induce angiogenesis in the mouse Matrigel plug assay. Thus, β-sitosterol is able to induce angiogenesis directly and/or indirectly. To confirm the direct angiogenic property of β-sitosterol, *in vitro* endothelial cell migration and proliferation assays were performed. The *in vitro* wound migration assay showed that β-sitosterol stimulated the motility of HUVECs. However, β-sitosterol did not promote endothelial cell proliferation. These observations are reminiscent of an earlier report that saponin from Ginseng Radix rubra stimulated migration, but not proliferation of HUVECs (Morisaki et al. 1995)

In conclusion, *Aloe vera* gel induced angiogenesis in the CAM assay and its most effective compound is β-sitosterol. In the presence of heparin, this plant-derived sterol induced neovascularization in *in vivo* mouse Matrigel plug assay, and promoted the migration, but not proliferation, of HUVECs *in vitro*. Collectively, it is clear from this study that *Aloe vera* gel promotes angiogenesis *in vivo*, and that this angiogenic effect is mainly due to β-sitosterol. These results suggest that the wound healing effect of *Aloe vera* gel is at least in part due to the angiogenic activity of β-sitosterol.

**References**
1. Amento EP, Beck LS (1991) TGF-β and wound healing. *Ciba Found Symp* 157, 115-123.
2. Awad AB, von Holtz RL, Cone JP, Fink CS, Chen YC (1998) Beta-sitosterol inhibits growth of HT-29 human colon cancer cells by activating the sphingomyelin cycle. *Anticancer Res* 18, 471-3.
3. Castellot JJ Jr, Karnovsky MJ, Spiegelman BM (1982) Differentiation-dependent stimulation of neovascularization and endothelial cell chemotaxis by 3T3 adipocytes. *Proc. Natl. Acad. Sci. USA.* 79, 5597-5601.
4. Davis RH, Leitner MG, Russo JM, Byrne ME (1989) Anti-inflammatory activity of *Aloe vera* against a spectrum of irritants. *J Am Pod Med Assoc* 79, 263-276.
5. Davis RH, Kabbani JM, Maro NP (1987) *Aloe vera* and Wound healing. *J. Am Pod Med Assoc* 77, 165-169.
6. Dunphy JE (1974) Modern biochemical concepts on te healing of wounds: Would healing. Baltimore, Williams and Wilkins 22-31
7. Folkman J, Klagsbrun M (1987) Angiogenic factor *Science* 235, 442-447.
8. Hagiwara H, Shimonaka M, Morisaki M, Ikekawa N, Inada Y. (1984) Sitosterol-stimulative production of plasminogen activator in cultured endothelial cell from bovine carotid artery. *Thromb Res* 33, 363-70.
9. Hebda PA, Klingbeil CK, Abraham JA, Fiddes JC (1990) Basic fibroblast growth factor stimulation of epidermal wound healing in pigs. *J Inv Dermatol* 95, 626-631.
10. Klein AD, Penneys NS (1988) *Aloe vera J Ame Aca Dermatol* 18, 714-720.
11. Lee MJ, Lee OH, Yoon SH, Lee SK, Chung MH, Park YI, Sung CK, Choi JS, Kim KW.(1998) *In vitro* angiogenic activity of *Aloe vera* gel on calf Pulmonary Artery endothelial (CPAE) cells. *Arch Pharm Res* 21, 260-5.
12. Lushbaugh CC, Hale DB (1953) Experimental acute radiodermatitis following beta irradiation. *Cancer* 6, 690-698.
13. McGee GS, Davidson JM, Buckley A, Sommer A, Woodward SC, Aquino AM, Barbour R, Demetriou AA (1988) Recombinant basic fibroblast growth Factor accelerates wound healing. *J Surg Res* 45, 145-153.
14. Moon EJ, Lee YM, Lee OH, Lee MJ, Lee SK, Chung MH, Park YI, Sung CK,Choi JS, Kim KW (1999) A novel angiogenic factor derived from *Aloe vera* gel: beta-sitosterol, a plant sterol. *Angiogenesis* 3,117-23.
15. Morisaki N, Watanabe S, Tezuka M, Zenibayashi M, Shiina R, Koyama N, Kanzaki T, Saito Y (1995) Mechanism of angiogenic effects of saponin from Ginseng *Radix rubra* in human umbilical vein endothelial cells. *Br J Pharmacol* 115, 1188-1193.
16. Phillips GD, Whitehead RA, Knighton DR (1991) Initiation and pattern of angiogenesis in wound healing in the rat. *Am J Anatomy* 192, 257-262.
17. Saiki I, Fujii H, Yoneda J, Abe F, Nakajima M, Tsuruo T, Azuma I (1993) Role of aminopeptidase N (CD13) in tumor-cell invasion and extracellular matrix degradation. *Int J Cancer* 54, 137-43.
18. Shimonaka M, Hagiwara H, Kojima S, Inada Y (1984) Successive study on the production of plasminogen activator in cultured endothelial cells by phytosterol. *Thromb Res* 36, 217-

19. Taylor S, Folkman J (1982) Protamine is an inhibitor of angiogenesis. *Nature* 297, 307-312.
20. Thompson WD, Harvey JA, Kazmi MA, Stout AJ (1991) Fibrinolysis and angiogenesis in wound healing. *J Pathol* 165, 311-8.
21. Tsuboi R, Rifkin DB (1990) Recombinant basic fibroblast growth factor stimulates wound healing in healing-impaired db/db mice *J Exp Med* 1990; 172, 245-251.
22. Weizel A, Richter WO (1997) Drug therapy of severe hypercholesterolemia. *Eur J Med Res* 2, 265-269.

## 4.3 Anti-allergic responses

**Ro, Jai Youl, Ph.D.**

Department of Pharmacology, Sungkyunkwan University School of Medicine

### *4.3.1. Introduction*

An immune response uses effector molecules that act by various mechanisms to eliminate antigen, which enters externally. These effector molecules induce a localized inflammatory response that eliminates antigen without severe damage to the host tissue. However, this inflammatory response can show in-appropriate effects under certain conditions, such as tissue damage or death.
This inappropriate immune response is called hypersensitivity or allergy.

Hypersensitivity reactions develop in the process of either cell-mediated or humoral immunity.

Hypersensitivity evoked from humoral responses is triggered by antibodies or by antigen/antibody reactions. This response is termed immediate hypersensitivity because the symptoms show within minutes or hours after a sensitized recipient encounters an antigen.

Gell P.G.H. and Coombs R.R.A. proposed a classification in which hypersensitivity reactions are divided into four types: I, II, III, and IV. Each type contains distinct mechanisms, cells, and mediator molecules (Abbas and Lichtman 2003; Goldsby et al. 2003). Here, we briefly describe the mechanisms of IgE-mediated hypersensitivity (type I), and then describe the anti-allergic effects and mechanisms of a single component, alprogen, extracted from *Aloe vera*.

Type I reactions may produce various events from severe life-threatening reactions, including asthma and systemic anaphylaxis, to hay fever and eczema, which are merely annoying. They are induced by antigens refer to as allergens.
They are divided into two classes: systemic anaphylaxis and localized anaphylaxis.

Systemic anaphylaxis is a shock-like state in which the onset of type I reaction occurs within minutes. In the guinea pig, the animal model that is chosen to study systemic anaphylaxis, a systemic reaction is easily induced with allergens, and its symptoms bear close similarity to those observed in humans. These allergens activate and degranulate mast cells. Therefore, it has been suggested that the extent of mast cell degranulation is a major determinant of clinical severity in allergens-induced, systemic anaphylaxis and mastocytosis.

Localized anaphylaxis (atopy) is limited to a specific target tissue or organ and is called atopy. Atopic allergies affect at least 20% of the population in developing countries and include allergic rhinitis (hay fever), asthma, atopic dermatitis (eczema), and food allergies caused by a wide range of IgE-mediated disorders.

Allergic rhinitis is the most common atopic disorder which is known as hay fever. Airborne allergens stimulate the sensitized mast cells in the conjunctivae and nasal mucosa and these cells induce the release of active mediators, which cause localized vasodilatation and increased capillary permeability.

Asthma includes two forms: extrinsic and intrinsic. The former (allergic asthma) is triggered by airborne or blood-borne allergens, such as dust, pollens, insect products or viral antigens. The latter (intrinsic) is induced by exercise or cold, etc.; they apparently differ from allergen stimulation. Asthma is triggered by mediators released from mast cell degranulation in the lower respiratory tract. Contraction of bronchial smooth muscles caused by mediators leads to bronchoconstriction, airway edema, mucus secretion, and inflammation, and then contributes to airway obstruction.

The asthmatic response is divided into early and late responses. The former occurs within minutes after allergen exposure and mainly involves histamine, leukotrienes ($LTC_4$, $LTD_4$), prostaglandin ($PGD_2$), and TNF-$\alpha$. Resulting effects lead to bronchoconstriction, vasodilatation, and production of mucus. The latter occurs hours later and involves various mediators, such as IL-4, IL-5, IL-13, IL-8, TNF-$\alpha$, INF-$\gamma$, platelet-activating factor (PAF), eosinophil chemotactic factor (ECF), and neutrophil chemotactic factor-anaphylaxis (NCF-A). These mediators increase endothelial cell adhesion and recruit inflammatory cells, especially eosinophils, mast cells and neutrophils, into the airway tissues. The recruited cells may induce tissue injury by releasing toxic enzymes, cytokines, and oxygen radicals, and then contribute to occlusion of the bronchial lumen with mucus, proteins, and cellular debris, edema, sloughing of the epithelium, thickening of the basement membrane, and hypertrophy of the bronchial smooth muscles.

Atopic dermatitis (allergic eczema), which is known as a skin inflammatory disease that is often associated with a family history of atopy, is frequently observed in young children. Serum IgE antibody levels are elevated. It has been reported that this dermatitis involves $Th_2$ cells in acute phases and $Th_1$ cells in chronic phases, and an increased number of eosinophils and mast cells.

Food allergies are caused by various food allergens. Allergen cross-linking of IgE on mast cells in the upper or lower gastrointestinal tract can induce localized smooth muscle contraction, increase of the permeability in mucous membranes.

IgE-mediated hypersensitivity reaction (Type I) is induced by various types of allergens. These allergens induce a humoral antibody response by the same mechanisms used for other soluble antigens, resulting in the generation of antibody-secreting plasma cells and memory cells. The difference of type I hypersensitivity reaction from a normal humoral response is that the B cells produce IgE antibody which binds with high affinity to Fc$\epsilon$ receptors on the surface of the mast cells or basophils. IgE-bound mast cells or basophils are referred to as sensitized cells. Second exposure to the same allergen cross-links and activates the mast cell or basophil membrane-bound IgE. These activated cells ultimately cause degranulation via various signal transductions. Effector cells associated with type I hypersensitivity are tissue mast cells and blood basophils. As mast cells were mainly used in this experiment, I am going to describe mainly mast cell activation

Mast cells have several features in common. Cells have surface receptors (FcεRI) that bind the Fc portion of IgE with high affinity. The cell surface-bound IgE cross-links with multi-antigens and secretes mediators that mediate allergic reactions. The mediators released by mast cells are either preformed and stored in secretory granules or newly generated.

The most common primary mediators (granular or preformed mediators) are histamine, proteases, ECF, NCF, and heparin. The secondary mediators are either released by the breakdown of membrane phospholipids or synthesized after target-cell activation. Newly generated mediators (secondary mediators) are produced in the mast cells activated by various stimuli. During the release process, arachidonic acid is released from cellular phospholipids via the activation of phospholipase enzymes. This arachidonic acid can then be metabolized along different pathways depending on the enzymes present, including the cyclooxygenase pathway, which produces prostaglandins (PGs), the lipoxygenase pathway, which metabolizes into leukotrienes (LTs), and the cytochrome $P_{450}$ pathway, which generates lipoxins. These mediators include LTs (LTB$_4$, LTC$_4$, LTD$_4$, and LTE$_4$), PGs (PGD$_2$), bradykinins, and PAF. The arachidonic acid is also metabolized by non-enzymatic pathway, yielding isoprostanes. These mediators are pharmacologically active agents that act on local tissues as well as on populations of secretory effector cells, including eosinophils, neutrophils, T lymphocytes, and platelets. However, the mediators of type I hypersensitivity partly differ among the various species or tissues.

Activation of mast cells also increases RNA expression and the release of several cytokines, such as IL-1, IL-3, IL-4, IL-5, IL-6, IL-8, IL-13, INF-γ, GM-CSF, and TNF-α, etc. Mast cells are also a source of growth and regulatory factors.

The initial event in the stimulation of the mast cells is the interaction of two IgE molecules on the cell surface with antigens. These interactions initiate a series of biochemical events that eventually result in the secretion of mediators from the cells. Therefore, the optimal conditions for the release of histamine depend on the concentration of antigen-specific IgE antibodies on the membrane, the affinity of the IgE for the antigen, and the concentration of the antigens. The increased intracellular calcium is also required in the IgE receptor-mediated release reaction. Aggregation of FcεRI receptors results in many biochemical events in the mast cells. As FcεRI does not have any intrinsic enzymatic activity, the activation of cellular biochemical events is due to the interaction of the aggregated receptors with other cytoplasmic or membrane proteins. The FcεRI receptor is a tetramer containing a α chain, a β chain, and a homodimer of disulfide-linked γ chains. The extracellular domain of the α chain of FcεRI binds IgE and its relatively short cytoplasmic domain probably does not play a role in cell signaling. In contrast, the COOH-terminal cytoplasmic domains of both β and γ subunits are important in FcεRI –mediated signal transduction. The cytoplasmic domains of β and γ subunits of FcεRI contain a motif that is important for signal transduction. This motif is called ITAM (immunoreceptor tyrosine-based activation motif; Asp/Glu-X$_2$-**Tyr**-X$_2$-Leu-X$_{6-8}$-**Tyr**-X$_2$-Leu/Iso, where X refers to any amino acid, and **Tyr** residues are phosphorylated in activated cells). However, various experiments demonstrate that the essential receptor subunit for signaling cells for secretion appears to be the γ subunit, and that molecules besides FcεRI are present in mast cell activation and play a critical role for propagating intracellular signals (Siragainan 1997; Gould et al. 2001; Abbas and Lichtman 2003; Goldsby et al. 2003).

If mast cells surface-bound IgE cross-link with multi-antigens, cells are activated. The activated cells secrete various types of mediators, which induce allergic reactions via multi-signaling pathway. Here, we describe the results based on experimental data indicating that alprogen, a single component purified from *Aloe vera*, exerts an anti-allergic effect in the mast cells activated with antigen/antibody reactions or in the human mast cell line (HMC-1) activation, and that NY945, the pre-step fraction of alprogen, has protective effects in animal models.

### 4.3.2. Anti-inflammatory and other effects of Aloe vera *extract*

*Aloe vera* is a plant medicine commonly used in basic health care in many countries including Korea. It has a history of medicinal use much longer than the last century. There have been many reports about the medicinal effect of *Aloe vera* according to development of modern natural science, especially on the gastrointestinal tract or immune system, etc. As seen ancient documents, *Aloe vera* has also been used to treat allergies and bronchial asthma. One of the fairly well documented preparations in traditional medicine is the whole or parenchyma's leaf-gel of *Aloe vera*. *Aloe vera* is a complex plant containing many biologically active substances. It has been reported that glycoproteins or sterols extracted from *Aloevera* have a strong anti-inflammatory response, and that polysaccharides in *Aloevera,* especially mannose-6-phosphate, have a strong wound healing activity and anti-inflammatory response (Davis 1994). Recently, it has been reported that acemannan, the main carbohydrate fraction extracted from parenchyma's leaf-gel of *Aloe vera*, activates the mouse macrophage cell line, induces the production of nitric oxide (NO), and regulates immune responses in various cells Zhang and Tizard 1998; Karaca 1995). It has been further reported that aloe-emodin inhibits neuroectodermal tumors, and that *Aloe vera* exhibits wound healing activity in second degree, rat burn model and induces the production of anti-oxidants (SOD, catalase) and immunostimulatory activity, etc (Williams et al. 1996). However, there are as yet no published reports other than ours about the effect of *Aloe vera* extract on the allergic reaction or asthmatic disorders.

On the basis of this literature review, it can be inferred that the effect of aloe extracts on inflammation may have therapeutic relevance to allergic hypersensitivity and asthmatic disorders. Therefore, we first attempted to purify crude *Aloe vera,* before examining whether the isolated single component of *Aloe vera*, alprogen, inhibits the mediator releases from guinea pig lung mast cells activated by specific antigen-antibody reactions, and also examining the mechanism of aloe extract. We also investigated the effect of NY945, the pre-step fraction of alprogen, on the guinea pig or mice allergic model. The purification of alprogen was performed in the laboratory of Professor Park, Young Ji as described in the section on aloe components.

### 4.3.2.1. Anti-allergic effect of *Aloe vera* extracts

**1. Effects of *Aloe vera* extracts on the guinea pig lung mast cell activation**

We examined whether S4301A, P4401T, G4201P, or S4101, various fractions extracted from fresh *Aloe vera*-leaf or dried powder, inhibited the release of histamine in guinea pig lung mast cells activated with specific antigen/antibody reaction. The glycoprotein fraction, P4401T, inhibited histamine release more strongly than those of the other fractions. Therefore, we examined whether the WIP (P4401T) fraction may have an effect on specific ovalbumin (OVA)-induced systemic anaphylactic shock in mice. All OVA-induced mice (positive control group) died after the induction of systemic anaphylaxis. However, mice pre-injected with WIP (1 mg/kg, i.p. injection) induced moderate anaphylactic shock in three out of 8 mice, and all the mice pre-injected with WIP (3 mg/kg) were found to be protected against anaphylaxis. Based on these results, single component alprogen and pre-step fraction NY945, which have anti-allergic response, were purified using general chemicals, adsorption, ion exchange column, and gel chromatography.

**1) Effect of WIPIGT or WIPIGC, glycoproteins extracted from *Aloe vera*, on the histamine release**

We examined whether each fraction extracted from *Aloe vera* has any inhibitory effect in guinea pig lung mast cells (4 x $10^5$ cells) activated with OVA/anti-OVA reaction. Mast cells were purified from guinea pig lung using enzyme digestion and Percoll gradient (Ro et al. 2000). The purified mast cells (4 x $10^5$ cells) were incubated with anti-OVA antibody (1ml/1 x $10^6$ cells) and challenged with OVA (0.1μg/ml). Each fraction extracted from *Aloe vera* was added 5min before OVA challenge. W1P1GT (50μg), the next fraction extracted from W1P, weakly inhibited the histamine release in guinea pig lung mast cells activated with OVA/anti-OVA reaction, whereas W1P1GT (100μg) significantly inhibited the histamine release during mast cell activation. The next fraction extracted from W1P1GT, W1P1GC (50μg), inhibited histamine release by 60% during mast cell activation.

**2) Effect of glycoprotein fraction extracted from *Aloe vera* dried powder on the histamine release**

In pretreatment with glycoprotein G2PGCP1-3 (50μg) or G2PGCP4 (50μg), histamine release was a 17.5% and 28.0% decrease, respectively, compared to the 33.2% of OVA alone. This result was not different from that of G2 1M PPT, which is a precipitate of *Aloe vera* dried powder. Glycoprotein from dried powder also inhibited histamine release more weakly than that of fresh *Aloe leaf*-leaf. It may be inferred that the active component in *Aloe leaf* is probably decomposed in the drying process. Therefore, we used fresh *Aloe leaf*-leaf for purification of the anti-allergic component.

### 3) Effect of three peaks of W1P1GC (aloe glycoprotein) on histamine release during mast cell activation

W1P1GC was eluted using a Sephacryl HR S200 column and then yielded three peaks. We examined the effect of each peak on the histamine release.

#### (1) Inhibitory effect of histamine release

When guinea pig lung mast cells activated with anti-OVA were challenged with OVA antigen, histamine release by W1P1GC (NY945), $P_1$, $P_2$, or $P_3$ (each 10μg) pretreatment were $8.1 \pm 1.26\%$, $8.7 \pm 2.07\%$, $10 \pm 1.89\%$ and $11.2 \pm 2.39\%$, respectively. These inhibitions were decreased by 69.4%, 67.2%, 58.9%, and 57.7% compared to the $26.2 \pm 2.21\%$ of OVA alone, respectively. Pretreatment with 30μg of each fraction inhibited histamine release by 90% in mast cell activation.

#### (2) Inhibitory effect of leukotriene synthesis

Leukotriene ($LTD_4$), a newly synthesized mediator released from mast cells activated with OVA/anti-OVA reaction, was measured by radioimmunoassay (RIA). Accounts of $LTD_4$ secreted by W1P1GC, P1, P2, or P3 (each 10μg) pretreatment were $76.2 \pm 2.53$, $62.1 \pm 3.26$, $58.1 \pm 4.91$, and $76.3 \pm 2.90$ pmole/$10^6$ cells, respectively. These inhibitions were decreased by 28.9%, 42.1%, 45.8%, and 29.0%, respectively, compared to the $107.2 \pm 1.21$ pmole/$10^6$ cells of OVA alone. P1 and P2 (30μg) inhibited $LTD_4$ by 60%, and P3 (50μg) by 35%.

#### (3) Inhibitory effect of tumor necrosis α (TNFα)

We could not measure directly guinea pig TNFα secreted from mast cells activated with antigen/antibody reaction because guinea pig TNFα and anti-TNF-α antibody were commercialized. Therefore, we measured the cross-linking activity between guinea pig and human TNFα using a human TNFα ELISA kit. The TNFα level of W1P1GC, $P_1$, $P_2$, or $P_3$ pretreatment was $0.418 \pm 0.098$, $0.765 \pm 0.101$, $0.983 \pm 0.248$, and $0.613 \pm 0.089$ pg/$10^6$ cells, respectively, and all levels were significantly decreased compared to that of OVA alone ($1.573 \pm 0.154$ pg/$10^6$ cells). The amount of TNFα secreted from guinea pig lung mast cells was smaller than that of rat peritoneal mast cells. However, it can be inferred that aloe extract inhibits TNFα secretion during guinea pig lung mast cell activation.

All peak $P_1$, $P_2$, or $P_3$ isolated from W1P1GC showed a similar degree of inhibition of histamine release by W1P1GC. Therefore, these data suggest that aloe glycoprotein products cleaved by some enzymes have better anti-histaminergic effects than fractionation of glycoprotein. Therefore, the effects of each fraction extracted by various glycosylase enzymes were examined in guinea pig lung mast cell activation.

### 4) Inhibitory effect of glycosyl chain part on histamine release

In order to examine the effect of glycosyl chain on histamine release, aloe glycoprotein was cleaved by peptide cleavage enzyme, yielding a glycosyl chain. When mast cells activated with OVA/anti-OVA antibody reaction were pretreated by the glycosyl chain fraction, histamine inhibited less than that achieved with NY945 pretreatment.

### 5) Inhibitory effect of protein part on histamine release

In order to examine the effect of protein part on histamine release, the glycosyl chain in aloe glycoprotein (NY945) was cleaved by N-glycosidase or O-glycosidase, yielding various protein parts which inhibited histamine release by 55% ~ 65%. These results were similar to the inhibition achieved with NY945. Therefore, it can be inferred from these data that only the glycosyl chain or NY945 peptide does not have any anti-histaminergic effect.

### 6) Inhibitory effect of alprogen, aloe single component, on the mediator release

The single components purified from fresh aloe-leaf were nominated as alprogen, alprogen I, and alprogen II, according to the purification method. Alprogen had the strongest anti-allergic effects (Table 1). Therefore, we mainly present below a description of the alprogen results.

## 2. Inhibitory effect of alprogen on the human umbilical cord blood-derived mast cell (HUCBMC) activation

Human umbilical cord blood-derived mast cells (HUCBMC) were cultured for 9~12 weeks using various growth factors such as IL-6, SCF and IL-3, etc. When mature HUCBMC cells ($2 \times 10^5$ cells) were sensitized and challenged with human anti-IgE (hIgE) antibody and human IgE antibody (hIgE) after allergen pre-treatment (5 min), histamine release decreased by 34.6% (from $16.2 \pm 1.6\%$ to $10.6 \pm 0.9\%$), and the increased IL-4 production ($8.50 \pm 0.41$pmol/$10^6$ cells) was reduced by $1.93 \pm 0.30$pg/$10^6$ cells. However, IL-13 was not secreted in the guinea pig lung mast cells activated with antigen/antibody reaction.

## 3. Inhibitory effect of alprogen on the mediator release and on cytokine production in the human mast cell line-1 (HMC-1)

We examined the effects of alprogen on the mediator release in HMC-1 cells activated with hIgE/anti-IgE reaction or with agonist, PMA.

### 1) Effect of alprogen on the mediator releases

Histamine release and LTD$_4$ production evoked by hIgE/anti-IgE-induced HMC-1 cells showed maximum amounts for 30min incubation of $32.4 \pm 2.79\%$ and $96.2 \pm 4.83$pmole/$10^6$ cells, respectively. Alprogen significantly inhibited the mediator releases in a dose-dependent manner. However, this significant inhibition was achieved in RPMI-culture media during mast cell activation, whereas HMC-1 cells did not release any mediators in IMDM culture media.

When HMC-1 cells ($1 \times 10^6$ cells) were stimulated with PMA (25ng) for 10 hrs, histamine release was $38.1 \pm 4.0\%$, but histamine was not released in IMDM culture media.

## 2) Effect of alprogen on the production of cytokines

When HMC-1 cells (1 x $10^6$ cells) were challenged by hIgE/anti-IgE antibodies, the production of GM-CSF or IL-4 was weakly increased. Alprogen pretreatment weakly, but not significantly, increased GM-CSF production, but it completely inhibited IL-4 production. Production of TNFα reached a maximum level (2,023.8 ± 33.8 pg/ml) in the HMC-1 cells activated with hIgE/anti-IgE anti-body reaction, but IL-8 production did not detect in HMC-1 cell activation.

Alprogen (5μg) reduced TNF-α production by 38.7% to 1,244.4 ± 35.1 pg/ml (Table 2).

When HMC-1 cells (1 x $10^6$ cells) were stimulated by PMA (25ng) for 3hr, the amounts of TNFα and IL-8 were remarkably increased to 1,667.7 ± 29.5 and 2,501.5 ± 40.2pg/ml, respectively. Alprogen (5μg) decreased TNFα and IL-8 to 833.9 ± 197.2 pg/ml and 1,300 ± 201.8pg/ml, respectively (Table 2). Based on the data, PMA-stimulated mast cells showed more cytokine production than antigen /antibody-challenged mast cells did. Therefore, the inhibitory mechanism of cyto-kines by alprogen was examined in the PMA-stimulated mast cell line.

## 4.3.2.2. Effect of NY945 or alprogen on the mediator releases and on signal pathways during guinea pig lung mast cell activation

Animal model and toxicity tests were performed by the pre-step fraction, NY945, because the amount of alprogen purified from aloe was very small (ap-proximately 2% of glycoproteins containing Aloe). Therefore, in order to examine the tests of animal model and toxicity, we compared the effect of NY945 to that of alprogen on the mediator releases and its mechanisms.

### 1. The effect of NY945 or alprogen on the mediator releases during mast cell activation

When the mast cells sensitized with anti-OVA antibody were challenged by 0.1 μg/ml OVA after alprogen II pretreatment of 2 μg/ml, histamine release was 21.6 ± 0.15%, which was 48.1% decreased compared to the 41.6 ± 0.10% of OVA alone. The amount of LTD$_4$ produced by 2 μg/ml alprogen II pretreatment was 48.0 ± 5.90 pmole/$10^6$ cells, which was 49.2% decreased compared to the 96.2 ± 4.83 pmole/$10^6$ cells of OVA alone (Table 1). In this study 5 μg alprogen pre-treatment completely inhibited both mediators released from the activated mast cells. The inhibitory effect of both mediator releases by alprogen pretreatment was dose-dependent (Ro 2000).

When mast cells were activated with calcium ionophore at 0.5 and 1.0μM, his-tamine release was 47.7 ± 1.5% and 62.6 ± 1.2%, and LTD$_4$ release was 39.6 ± 5.69 pmole/4x$10^5$ cells and 52.7 ± 8.56 pmole/4x$10^5$ cells, respectively. However, alprogen (5μg) did not inhibit either of the histamine and leukotriene releases evoked by calcium ionophore (0.5, 1.0μM). Therefore, these data suggest that alprogen selectively acts on mast cells activated with antigen/antibody reaction more than with agonists like calcium ionophore.

NY945 (20µg or 50µg) inhibited histamine release by 41.1% and 75%, and leukotriene (LTD$_4$) release by 23.7% and 46.8%, respectively (Table 1). These results were 5 ~ 10 times less than those of alprogen.

**Table 1.** Effect of NY945, Alprogen, and Alprogen I and II on the mediator releases from passively sensitized (anti-OVA) guinea pig lung mast cell activation caused by OVA.[a]

| Treatment | Histamine (%) | Leukotrienes (pmole/10$^6$ cells) |
|---|---|---|
| OVA alone | 41.6 ± 0.10 | 96.2 ± 4.83 |
| NY945 (µg) | | |
| 20 | 24.5 ± 2.09[*] (41.1)[b] | 73.4 ± 3.05 (23.7)[*] |
| 50 | 10.4 ± 1.02[***] (75.0) | 51.2 ± 5.11 (46.8)[**] |
| Alprogen (µg) | | |
| 2.0 | 33.3 ± 0.21[*] (20.0)[b] | 70.2 ± 3.71[*] (27.0) |
| 5.0 | 24.5 ± 1.15[**] (41.1) | 52.1 ± 4.11[**] (45.8) |
| Alprogen I (µg) | | |
| 2.0 | 34.9 ± 0.45[*] (16.1) | 82.3 ± 6.73 (14.4) |
| 5.0 | 28.7 ± 1.28[*] (31.0) | 69.3 ± 5.18[*] (28.0) |
| Alprogen II (µg) | | |
| 2.0 | 21.6 ± 0.15[***] (48.1) | 48.9 ± 5.90[**] (49.2) |
| 5.0 | 0.4 ± 0.01[***] (99.0) | 5.0 ± 5.00[**] (94.8) |

[a] Guinea pig lung mast cells were isolated and purified by digestion and Percoll density gradient method. Mast cells (0.4 x 10$^6$) were passively sensitized by anti-OVA antibody, and challenged by OVA (0.1µg /ml). Alprogen was added 5min before OVA challenge.
[b] Figures in parentheses are the decreasing percentage evoked by Alprogen pretreatment.
*, $P<0.05$; **, $P<0.01$; ***, $P<0.001$

**Table 2.** Effect of Alprogen on the production of cytokines in the human mast cell line (HMC-1) activated with hIgE/anti-IgE antibody or PMA[a].

| | Challenge Time | GM-CSF (pg/ml) | IL-4 (pg/ml) | TNFα (pg/ml) | IL-8 (pg/ml) |
|---|---|---|---|---|---|
| Anti-IgE | 30min | $0.1 \pm 0.03$ | $4.7 \pm 0.24$ | $765.7 \pm 53.00$ | -[b] |
| Alprogen (5μg) | 30min | $0.1 \pm 0.01$ | $4.4 \pm 0.34$ | $588.3 \pm 10.50^*$ | |
| Anti-IgE | 12hr | $0.5 \pm 0.34$ | $8.8 \pm 0.17$ | $1317.3 \pm 85.30$ | -[b] |
| Alprogen (5μg) | 12hr | $4.6 \pm 0.10^*$ | $4.9 \pm 0.59^*$ | $982.4 \pm 34.80^*$ | |
| Anti-IgE | 24hr | $1.0 \pm 0.11$ | $8.41 \pm 0.95$ | $2023.8 \pm 33.30$ | -[b] |
| Alprogen (5μg) | 24hr | $5.2 \pm 0.27$ | $5.1 \pm 0.40^*$ | $1244.4 \pm 35.10^*$ | |
| PMA (25 ng) | 3hr | -[b] | -[b] | $1667.7 \pm 29.50$ | $2501.5 \pm 40.20$ |
| NY945 (50μg) | 3hr | -[b*] | -[b] | $833.9 \pm 197.20^*$ | $1300.8 \pm 201.80^*$ |

[a] HMC-1 (4 x $10^5$ cells) were sensitized with myeloma human IgE (1.0μg/ml) for 5hr at 37°C, and challenged with recombinant human anti-IgE antibody (1.0μg/ml) or PMA.
[b] No detection; *, P<0.01 by comparison without Alprogen

## 2. The effect of NY945 or alprogen on the influx of $Ca^{2+}$ during mast cell activation

It has been reported that the increase of intracellular $Ca^{2+}$ ($[Ca^{2+}]_i$) and the activation of PKC are necessary for degranulation of preformed and newly synthesized inflammatory mediators in mast cells (Altrichter et al. 1995). Therefore, the effects of NY945 or alprogen on $[Ca^{2+}]_i$ were examined in a single mast cell and visualized with confocal laser scanning microscopy by fluorescence intensity (optical density, O.D.) via fixation of Fluo-3 (5μM). $[Ca^{2+}]_i$ in a single mast cell reached a plateau at 14 sec (from 20,022 ± 75 to 25,291 ± 951) in the presence of external $Ca^{2+}$ (Fig. 1). Furthermore, $[Ca^{2+}]_i$ was significantly decreased by NY945 (10μg) or alprogen (1μg) in a dose dependent manner (from 20,033 ± 156 to 19,633 ± 630 for 1μg; from 20,191 ± 90 to 17,517 ± 762 for 5μg at 14 sec). Based on these results, it can be inferred that NY945 or alprogen acts on multi-signal transduction via the inhibition of $Ca^{2+}$ influx. This suggestion is based on the activities of enzymes or 2nd messengers that are related to signal pathways being $Ca^{2+}$-dependent events.

It is therefore essential that intracellular $Ca^{2+}$ concentration be increased for mediator releases.

## 3. The effect of NY945 or alprogen on the activation of the phospholipase D activity during mast cell activation

When the sensitized mast cells are activated with antigen (antigen/antibody reaction), two enzyme pathways, phospholipase C (PLC) and phospholipase D (PLD), are activated, after which they produce DAG (diacylglycerol). Recently, it has been demonstrated that a number of DAG arising during mast cell activation was indirectly produced via PLD enzyme activation. That is, an increase of membranous PLD activity during the mast cell activation evoked by specific antigen-antibody reactions ultimately leads to the release of mediators from mast cells.

Therefore, we studied the effects of NY945 or alprogen on the increasing PLD activity in mast cells activated by OVA-anti-OVA antibody reactions. The PLD activity was measured by phosphatidylbutanol (PBut) produced in the presence of an aliphatic alcohol (butanol).

The production of PBut in the activated mast cell increased remarkably from 3,237 ± 669 cpm to 11,555 ± 570 cpm, but with alprogen (0.5μg) pretreatment it was deceased remarkably from 11,555 ± 570 cpm to 6,592 ± 659 cpm. The PLD activity following alprogen pretreatment was decreased by 43% when compared to antigen alone and this decrease occurred in a dose-dependent manner. However, the PLD activity by NY945 exhibited a level 5 ~ 10 fold less than that of alprogen.

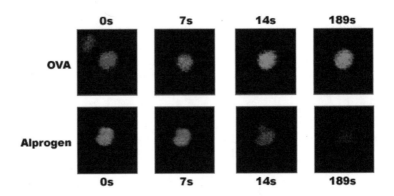

**Fig. 1.** Effect of Alprogen on the $Ca^{2+}$ influx in the mast cells activated with OVA/anti-OVA reaction. The purified mast cells ($2\times10^5$ cells) were sensitized and challenged with anti-OVA and OVA, respectively. Fluo-3, AM (5.0uM) was added to cells which were fixed on slide glass and activated by 0.1ug/ml OVA in the absence or presence of alprogen (2μg). OVA, OVA challenge; s, second. (See Plate 11.)

### 4. Effect of NY945 or alprogen on PLC activity during mast cell activation

In the presence of excess butanol (100mM), DAG produced by PLD activation was completely inhibited in mast cells activated with antigen/antibody reaction, after which the amounts of DAG generated were produced by PLC activity. The generated DAG was also increased from $1,192 \pm 152$ cpm to $15,787 \pm 952$ cpm. In mast cell activation, NY945 (10μg) pretreatment reduced the DAG produced by PLC activity from $15,787 \pm 952$ cpm to $3,636, \pm 928$ cpm. Alprogen (2μg) completely inhibited DAG production. Therefore, these data suggest that NY945 or alprogen may inhibit PLC enzyme activity in guinea pig lung mast cells activated with OVA/anti-OVA reaction.

### 5. The effect of NY945 or alprogen on the production of mass 1, 2- diacylglycerol during mast cell activation

PKC activity is activated by DAG. The mass DAG is formed by the activation of phospholipases such as PLD and PLC in lung mast cells activated by OVA-anti-OVA reactions (Kettner et al. 2004). Therefore, we investigated the effect of NY945 or alprogen on the formation of mass DAG in the lung mast cells activated by OVA/anti-OVA reactions. From $36,200 \pm 900$ cpm, mass DAG production decreased by 58.3% to $15,111 \pm 389$ cpm after the treatment with NY945 (10μg) and by 39.5% to $21,900 \pm 300$ cpm after alprogen treatment (2μg). In the treatment of alprogen (5μg), the formation of mass DAG was completely blocked. Therefore, inhibition of DAG by alprogen was 7 fold stronger than that by NY945.

**6. Effect of NY945 or alprogen on the adenylate cyclase activity during mast cell activation**

It has been known that mast cells activated with antigen/antibody reaction transiently increased intracellular cAMP level via activation of adenylate cyclase, followed by degranulation of mast cells and finally release of mediators. NY945 (10 $\mu$g) and alprogen did not have any inhibitory effect on adenylate cyclase (data not shown).

**7. The effect of NY945 or alprogen on protein kinase C (PKC) activity during mast cell activation**

Since $[Ca^{2+}]_i$ is decreased by the treatment of NY945 or alprogen, it may also influence PKC activity (Kettner et al. 2004). Phosphorylated protein was monitored in order to measure the effects of NY945 or alprogen on PKC activity. Phosphorylated proteins activated by OVA/anti-OVA reactions were increased by approximately 7.8 times (from 27,793 ± 10,200 cpm to 217,755 ± 20,334 cpm). From 217,755 ± 20,334 cpm, NY945 (10$\mu$g) or alprogen (2$\mu$g or 5$\mu$g) decreased the activity of PKC by 35.1%, 37.4% or 51.3%, to 141,322 ± 8,325 cpm, 136,232 ± 9,432 cpm, and 106,118 ± 19,428 cpm, respectively.

**8. The effect of NY945 or alprogen on methyltransferase (MT) activity during mast cell activation**

When mast cells are activated with antigen/antibody cross-linking, MT is activated. This event causes methylation of phospholipids, followed by mediator releases. Therefore, we examined whether NY945 or alprogen inhibits MT enzyme activity in mast cell activation. From 17,475 ± 375 cpm, NY945 (10$\mu$g) and aiprogen (2$\mu$g) reduced MT activity by 33.5% and 45%, to 11,620 ± 437 cpm and 9,611 ± 346 11,620    437 cpm and 9,611 ± 346 cpm, respectively.

The methylated phospholipids separated by TLC were lyso-phosphatidylcholine(lyso-PC), phosphatidylcholine (PC), phosphatidyldimethylethanolamine (PDME), and phosphatidylmonomethylethanolamine (PMME). In particular, the first two were inhibited by NY945 or alprogen. Therefore, these data suggest that Aloe extract inhibits the production of arachidonic acid, precursor of leukotriene production, via PC inhibition which is known as a substrate for $PLA_2$ enzyme, and that lyso-PC reduced by Aloe extract inhibits the flexibility of cell membrane by inhibiting the formation of $Ca^{2+}$ channel or $Ca^{2+}$ influx, thereby inhibiting $Ca^{2+}$ -dependent $PLA_2$ activity.

**9. The effect of NY945 or alprogen on the activation of $PLA_2$ during mast cell activation**

The increase of $[Ca^{2+}]_i$ has a number of effects on mast cells, one of the major of which is the induction of the association of $PLA_2$ with membranes to facilitate the synthesis of lipid mediators such as leukotrienes and prostaglandins (Fontech et al. 2000). Since the release of arachidonic acid is partly responsible for Type I hypersensitivity reactions, we investigated whether NY945 or alprogen inhibits $PLA_2$ activity. The activity of $PLA_2$ increased by 41.0% in mast cells activated by 0.1$\mu$g /ml OVA challenge (from 8,237 ± 554 cpm to 11,617 ± 769 cpm). However, the activity of PLA$_2$ decreased by 20% or 26.7% (from 11,617 ± 769 cpm, to 9,293 ± 573 cpm for NY94, and to 8,520 ± 1,144 for alprogen) by treatment with NY945 (10$\mu$g) or alprogen (5$\mu$g), compared to that of OVA challenge alone. Therefore, these data suggest that NY945 or alprogen inhibits the production of leukotriene, which is known as an allergic mediator 1,000 fold stronger than histamine, via inhibition of PLA$_2$ activity.

### 4.3.2.3. Mechanism of cytokine production in PMA-induced HMC-1 cell line

HMC-1 cells stimulated with hIgE/anti-IgE antibody or PMA did not release mediators (histamine and LTD$_4$), but instead they secreted inflammatory cytokines such as IL-4, TNF-$\alpha$, IL-13, and IL-8. As the relationship between cytokine production and MAP kinases has been reported in human cord blood-derived mast cells or RBL-2H3 cells activated with antigen/antibody reaction, we examined the phosphorylation of MAP kinases. PMA-induced HMC-1 cells increased the phosphorylation of MAP kinases like ERK, p38, and JNK. NY945 (10$\mu$g) or alprogen (2$\mu$g) inhibited the phosphorylation of all MAP kinases (data not shown).

### 4.3.2.4. Effects of NY945 on the allergic rhinitis animal model

As described above, because NY945 and alprogen showed anti-allergic effects in vitro, we examined the effect of aloe extracts on allergic rhinitis and asthma in vivo. However, we could not use alprogen for animal model experiments because the amounts of alprogen purified from fresh *aloe vera* whole-leaf (2% of glycoprotein) are very small. So, NY945 that has an effect similar to that of alprogen was used for the animal experiments.

#### 1. Effect of NY945 on total airway conductance (TAC)

A guinea pig allergic rhinitis model was prepared by intraperitoneal (i.p.) injection of OVA. Five, 10, 15, 20, 25, and 30 min after injection, the relative conductance was gradually reduced to 0.92 ± 0.23, 0.73 ± 0.14, 0.57 ± 0.13, 0.45 ± 0.11, 0.49 ± 0.06, and 0.60 ± 0.04, respectively, and the final level at 30 min after injection showed the recovery stage. NY945 (5mg/kg) pre-injection recovered the reduced TAC (to 0.84 ± 0.41, 0.71 ± 0.40, 0.62 ± 0.33, 0.74 ± 0.45, 0.69 ± 0.44, and 0.71 ± 0.45, respectively), and the conductance was strongly increased 20 min after NY945 injection.

#### 2. Effect of NY945 on the total airway resistance (TAR)

In the allergic rhinitis guinea pig, relative resistance strongly increased 20 min after OVA injection (1.29 ± 0.20 at 5min, 1.36 ± 0.31 at 10min, 1.85 ±0.49 at 15min, 2.36 ± 0.47 at 20min, 2.01 ± 0.32 at 25min, and 1.70 ± 0.26 at 30min). NY945 (3mg/kg or 5mg/kg) strongly reduced the elevated TAR caused by OVA injection. TAR after 5mg/kg NY945 pre-injection was 0.79 ± 0.40, 0.95 ± 0.46, 1.10 ± 0.42, 0.94 ± 0.37, 0.94 ± 0.46, and 0.87 ± 0.47, respectively. When NY945 was injected 1 week before OVA-sensitized guinea pigs, 5mg/kg, but not 3mg/kg,

NY945 reduced TAR (Fig. 2). These data suggest that NY945 may have protective and therapeutic effects for the elevated TAR in allergic rhinitis.

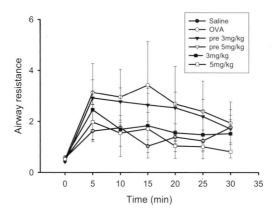

**Fig. 2.** Effect of NY945 on the increase of airway resistance induced by OVA in sensitized guinea pig. NY945 was injected intraperitoneally 1 week before antigen sensitization or 1hr before antigen challenge. Antigen was performed by 1% OVA nebulization for 10min. Total airway resistance was measured with single chamber plethysmography. Saline, negative control; OVA, OVA challenge; Pre-3mg/kg and pre 5mg/kg, 1week before antigen sensitization; 3mg/kg and 5mg/kg, 1h before antigen challenge.

### 3. Effect of NY945 on the allergic symptom score

In the allergic rhinitis guinea pig model, sneezing and rubbing scores were counted for 30 min after OVA challenge to give a total allergic symptom score of 43.5 ± 13.2. NY945 (3mg/kg or 5mg/kg) remarkably reduced this symptom score by 19.0 ± 8.2 and 21.8 ± 9.2, respectively. However, this reduction was not dose-dependant.

### 4. Effect of NY945 on the inflammatory cell recruitment in guinea pig nasal septal mucosa or lung and tracheal tissues

After the relative conductance and resistance in OVA-induced allergic rhinitis animals were measured, nasal septal mucosa, lung, and tracheal tissues were isolated, embedded in paraffin, and stained with Luna or toluidine blue stain, after which we observed the infiltrated inflammatory cells.

### 1) Effect of NY945 on the eosinophil infiltration in guinea pig mode

In nasal septic mucosa, lung and tracheal tissues of the guinea pig model, a number of eosinophils were observed in perivascular and peribronchiolar connective tissues. NY945 (3mg/kg or 5mg/kg) inhibited eosinophil infiltration dose-dependently (Fig. 3).

### 2) Effect of NY945 on the eosinophil infiltration in rat model

In nasal septic mucosa, lung and tracheal tissues of the rat model, eosinophils were observed in perivascular and peribronchiolar connective tissues, but the numbers were smaller than those in the guinea pig model. Eosinophil counts by NY945 (3mg/kg or 5mg/kg) pretreatment were reduced to 3.82 ± 5.97 cells and 2.19 ± 1.62 cells, respectively, compared to 11.64 ± 6.80 cells in a visible area.

### 3) Effect of NY945 on the mast cell infiltration in guinea pig model

Mast cells localized in nasal septic mucosa, lung, and tracheal tissues were examined using 0.1% toluidine blue stain. Mast cells were bigger than eosinophils and violet blue-colored cytoplasmic cells were observed in the epithelial layer. NY945 (3 mg/kg or 5mg/kg) significantly reduced mast cells infiltrated into all three tissues (Fig. 4).

## Plate 1

*Aloe barbadensis Mill   Aloe arborescens Mill   Aloe saponaria (Ait.) Haw*

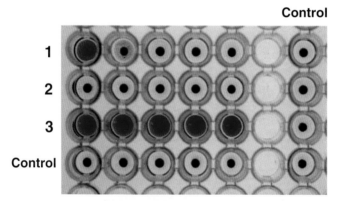

**Fig. 2.** Visual assay on microtiter plates of the hemagglutinating activity of peaks 1 to 3 purified on Sephacryl S-300 HR gel filtration, using rabbit erythrocytes. Each well contained 0.05ml of erythrocyte suspension and the first well contained 0.05ml of peak solution, with serial two-fold dilutions in adjoining wells. The control contained only rabbit erythrocyte suspension and 20 mM Tris.Cl, pH 7.4.

**Plate 2**

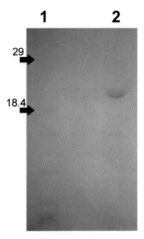

**Fig. 3.** The pattern of 16.5% SDS-PAGE of aloe lectin purified by glucose-affinity column chromatography. The arrows in Lanes 1 and 2 represent a size marker and aloe lectin, respectively.

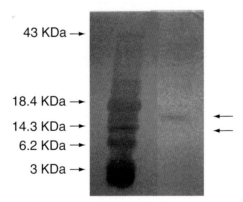

**Fig. 7.** The band pattern of *Aloe vera* proteinase inhibitor separated on 12.5% SDS-PAGE

# Plate 3

**STI   AVPI   Trypsin**

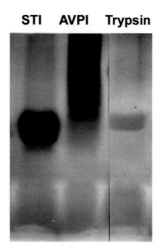

**Fig. 8.** Reverse zymography of AVPI (150 μg of Soybean trypsin inhibitor, 100 μg of *Aloe vera* proteinase inhibitor, 150 μg of Trypsin).

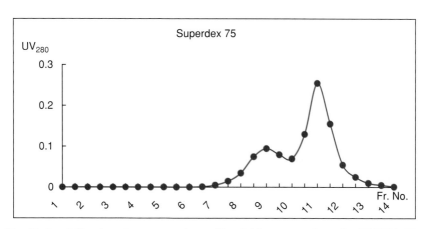

**Fig. 11.** A gel filtration chromatography profile of Alprogen on Superdex 75 10/30. The estimated volume of Alprogen is 10.1 ml. The arrow indicates the alprogen peak.

## Plate 4

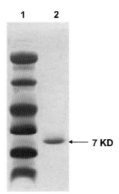

**Fig. 12.** Protein analysis of Alprogen by 16.5% SDS-PAGE. The gel electrophoresis was carried out through 16.5%(w/v) polyacrylamide gel containing 0.1% SDS. Lane 1 is a size marker, Lane 2 is Alprogen.

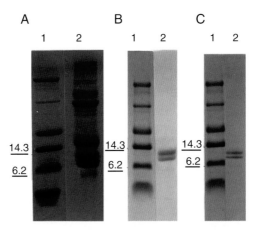

**Fig. 13.** Protein analysis of the NY945 and Alprogens by 16.5% SDS-PAGE. The gel electrophoresis was carried out through 16.5% (w/v) polyacrylamide gel containing 0.1% SDS. A: lane 1, protein size marker; lane 2, NY945. B: lane 1, protein size marker; lane 2, Alprogen I. C: lane 1, protein size marker; lane 2, Alprogen II.

Plate 5

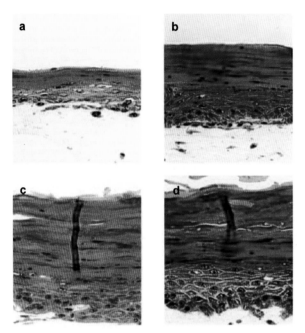

**Fig. 8.** Effect of glycoprotein fraction G1G1M1DI2 on the formation of epidermis from keratinocytes in raft culture. Cultures were treated with G1G1M1DI2 for 21 days immediately after exposure to the air. Cultures were grown (a) in the absence of G1G1M1DI2, or in the presence of G1G1M1DI2 at a concentration of (b) 0.05, (c) 0.5 or (d) 50 ug/ml (haematoxylin and eosin, original magnification×200).

# Plate 6

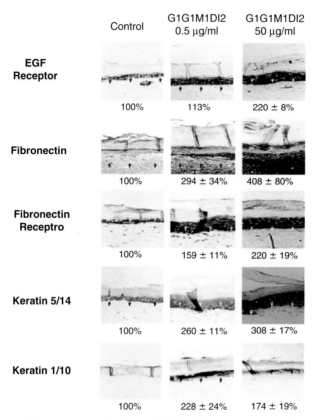

**Fig. 9.** Immunohistochemistry of the artificial epidermis raised in the raft culture. Immunohistochemical staining was used to detect epidermal growth factor (EGF) receptor, fibronectin, fibronectin receptor, keratin 5/14 and keratin 1/10 in the artificial epidermis raised in the absence or presence of 0.5 to 50 ug/ml G1H1M1DI2. Immunostaining is indicated by arrowheads. Data below each picture represent mean ± SD of relative percentage of optical density of three experiments. The immunostaining intensity was estimated by an image analyzer with image analysis software (immunogold staining, original magnification × 200).

Plate 7

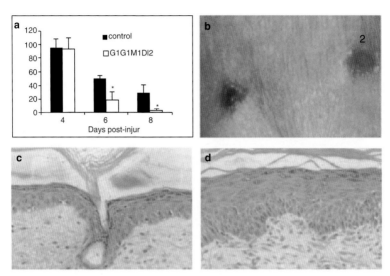

**Fig. 10.** Effect of glycoprotein fraction G1G1M1DI2 on the wound healing of hairless mice. (a) Data represent relative percentage of wound size (mean±SD; n=10). The wound area was photographed and the size of the wound was estimated using a BAS-2500 image analyser with image analysis software MCID ver. 3.0 and expressed as a percentage of wound size at day 0 after injury. *P<0.05 compared with control (Student's t-test). (b) Wound area on day 10 after treatment: 1, control; 2, G1G1M1DI2-treated group (10 mg/g ointment). (c, d) Morphology of the wound area at day 10 after treatment: (c) control (d) G1G1M1DI2-treated group (haematoxylin and eosin, original magnification × 200).

**Plate 8**

(a)     Blank coverslip          (b)     PMA (120 ng/egg)

(c)     G1 (500 µg/egg)          (d)  β-sitosterol (10 µg/egg)

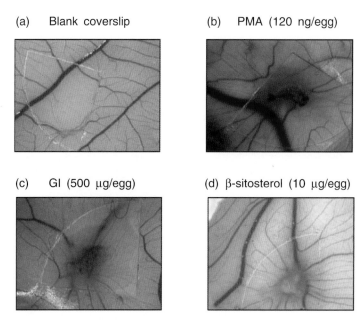

**Fig. 3.** Effects of G1 and β-sitosterol on embryonic angiogenesis in CAM 3 after sample implantation. Fat emulsion (10%) was injected into chorioallantois to clear the vascular network. (a) Control CAMs treated with blank coverslips showed no angiogenic response. Positive responses were seen in CAMs implanted with coverslips loaded with (b) PMA (120 ng/egg), (c) G1 (500 µg/egg), and (d) β-sitosterol (10 µg/egg).

Plate 9

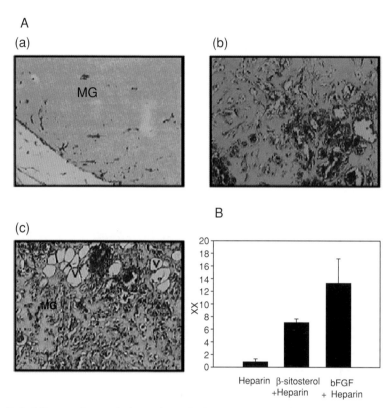

A

(a) (b)

MG

B

(c)

**Fig 5.** A. Microscopic morphology of Matrigel plugs. Masson Trichrome-stained Matrigel plug impregnated with heparin (a); Masson Trichrome-stained Matrigel plus impregnated with β-sitosterol plus heparin (b); and Matrigel plug impregnated with bFGF plus heparin, as positive control (c). Male C57BL/6 mice were injected subcutaneously with either Matrigel containing heparin alone (40 unit/ml), heparin plus β-sitosterol (30 μg/ml) or heparin plus bFGF (100 ng/ml). Plugs were removed 5 days after injection and processed for histology. Magnification (100 X). B. Hemoglobin (Hb) content was measured by the Drabkin method using Drabkin reagent kit 525 (Sigma, St Louis, MO). The concentration of Hb was calculated from a known amount of Hb assayed in parallel. MG, Matrigel; v, vessels.

**Plate 10**

**A**

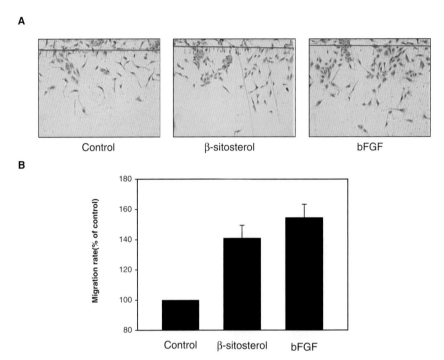

**B**

**Fig. 6.** Effects of β-sitosterol on the migration of HUVECs. HUVECs were seeded on a gelatin-coated culture dish. (A) At the 90% confluency, the monolayers were wounded with a razor blade. Wounded cells were rinsed with serum free medium and then incubated with β-sitosterol (2 μM) for 24h. bFGF (30 ng/ml) treated HUVECs were used as a positive control. The number of HUVECs that moved beyond the reference line was counted. (B) Results are expressed as percentages of control ± SEM. *, $P < 0.001$ compared with the control.

# Plate 11

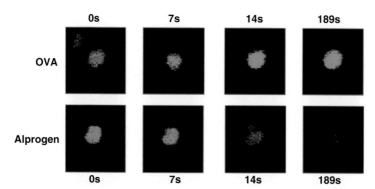

**Fig. 1.** Effect of Alprogen on the Ca2+ influx in the mast cells activated with OVA/anti-OVA reaction. The purified mast cells ($2 \times 105$ cells) were sensitized and challenged with anti-OVA and OVA, respectively. Fluo-3, AM (5.0uM) was added to cells which were fixed on slide glass and activated by 0.1 ug/ml OVA in the absence or presence of alprogen (2μg). OVA, OVA challenge; s, second.

# Plate 12

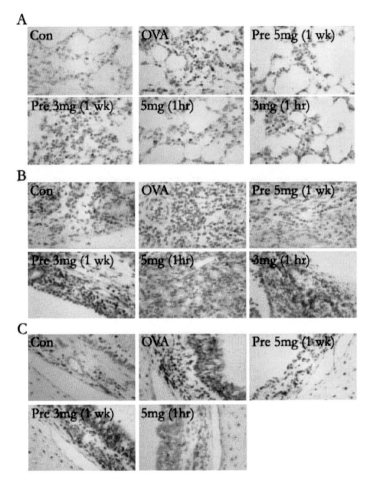

**Fig. 3.** Photograph (× 400, Luna stain) of lung (A), nasal mucosa (B), and trachea (C) in activated guinea pigs. NY945 was injected i.p. on 1, 3, 5, and 7 days before general sensitization or 1hr before antigen challenge. Antigen was performed by 0.1% OVA nebulization for 10min. After airway resistance measurement, each tissue was stained with Luna dye to examine the eosinophil infiltration. Con, negative control; OVA, OVA challenge; Pre-3mg/kg and pre 5mg/kg, 1 week before antigen sensitization; 3mg/kg and 5mg/kg, 1h before antigen challenge.

Plate 13

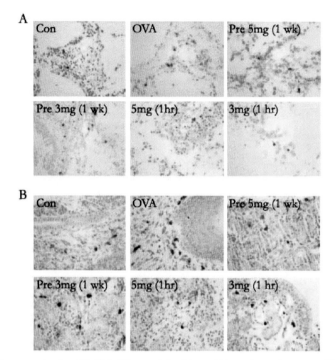

**Fig. 4.** Photography ( × 400, Toluidine stain) of lung (A), and nasal mucosa (B), in activated guinea pigs. NY945 was injected i.p. on 1, 3, 5, and 7 days before general sensitization or 1hr before antigen challenge. Antigen was performed by 0.1% OVA nebulization for 10min. After airway resistance measurement, each tissue was stained with toluidine dye to examine the mast cell infiltration. Con, negative control; OVA, OVA challenge; Pre-3mg/kg and pre 5mg/kg, 1 week before antigen sensitization; 3mg/kg and 5mg/kg, 1h before antigen challenge.

## Plate 14

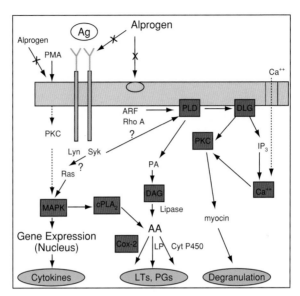

**Fig. 5.** Possible action site of Alprogen and pathway of signal transduction in mast cell activation.

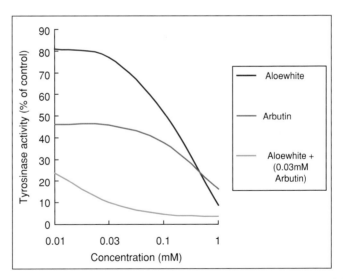

**Fig. 3.** Dose-dependent inhibition of aloewhite and arbutin on the tyrosinase activity. Each data is the mean of triplicate determinations ± SE. The tyrosinase activity was determined by Pomerantz method.

**Plate 15**

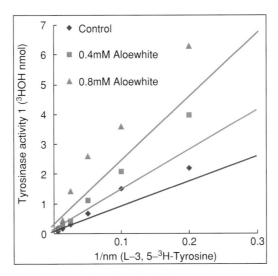

**Fig. 4.** Kinetics of tyrosinase inhibition by Aloewhite.

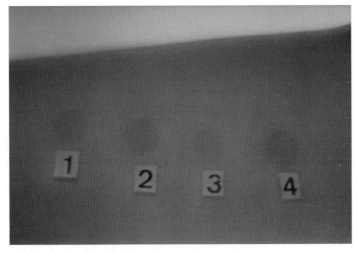

**Fig. 5.** *In vivo* assay for the whitening effect on human skin.
1. negative control (++++)
2. Aloewhite administered (++)
3. .Aloewhite + Arbutin administered (-)
4. Arbutin administered (++++)
5. (+ or - represent relative color intensity)

## Plate 16

0    0.01   0.1   1   5   25   50   (µm)

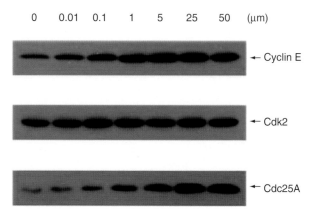

← Cyclin E

← Cdk2

← Cdc25A

**Fig. 4.** Effects of aloesin (A) and methylaloesin (B) on the expression of cyclin E, cdk2 and cdc25A proteins in SK-HEP-1 cells: SK-HEP-1 cells were cultured in medium with increasing concentration of aloesin or methylaloesin for 24h. Cell lysates were prepared and subjected to 12% SDS-PAGE, western blotting, and immunodetection. Cyclin E, cdk2, and cdc25A proteins were detected by polyclonal anti-cyclin E, cdk2, and cdc25A antibodies, respectively, using ECL method.

0    0.01   0.1   1   5   25   50   (µm)

← Cdc25A

**Fig. 5.** Effects of aloesin and methylaloesin on cdk2 kinase activity in SK-HEP-1 cells: SK-HEP-1 cells were cultured in medium with increasing concentration of aloesin or methylaloesin for 24 h. Cell lysates were prepared and immune-complex kinase assays were performed using histone H1 as substrate.

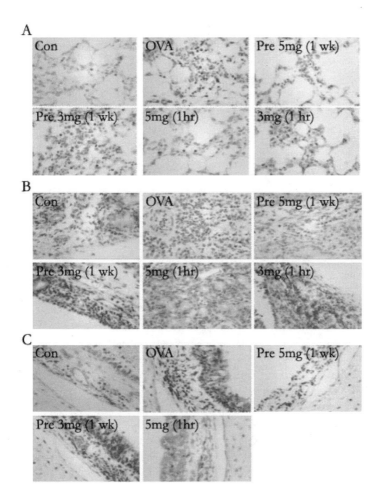

**Fig. 3.** Photograph (x 400, Luna stain) of lung (A), nasal mucosa (B), and trachea (C) in activated guinea pigs. NY945 was injected i.p. on 1, 3, 5, and 7 days before general sensitization or 1hr before antigen challenge. Antigen was performed by 0.1% OVA nebulization for 10min. After airway resistance measurement, each tissue was stained with Luna dye to examine the eosinophil infiltration. Con, negative control; OVA, OVA challenge; Pre-3mg/kg and pre 5mg/kg, 1week before antigen sensitization; 3mg/kg and 5mg/kg, 1h before antigen challenge. (See Plate 12.)

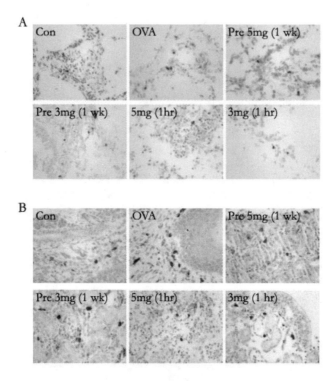

**Fig. 4.** Photograph (x 400, Toluidine stain) of lung (A), and nasal mucosa (B), in activated guinea pigs. NY945 was injected i.p. on 1, 3, 5, and 7 days before general sensitization or 1hr before antigen challenge. Antigen was performed by 0.1% OVA nebulization for 10min. After airway resistance measurement, each tissue was stained with toluidine dye to examine the mast cell infiltration. Con, negative control; OVA, OVA challenge; Pre-3mg/kg and pre 5mg/kg, 1week before antigen sensitization; 3mg/kg and 5mg/kg, 1h before antigen challenge. (See Plate 13.)

### 4.3.2.5. Toxicity test

General pharmacological test, acute and sub-acute tests were performed following the standard of "Toxicity Test" according to the presentation number 96-8(1996. 4. 16.) of the Korean Food & Drug Administration. As the aloe extract for the toxicity test, NY945 was used instead of alprogen because of the requirement for many samples.

### 1. Acute toxicity test (Lethal dose, $LD_{50}$)

NY945 was administrated with a single i.p. injection of 0.5mg/kg to 5mg/kg in male or female ICR mice or Sprague Dawley rats. However, but all mice and rat did not die (N = 10 animals/group). Therefore, we assumed that the $LD_{50}$ of NY945 is above 5mg/kg. We did not continue the experiment at higher NY945 doses because of the long preparation time for purification of NY945.

### 2. General pharmacological test

NY945 (40µg, 200µg, 1mg/kg, or 3mg/kg) was injected by i.p. in mice or rats, after which we observed the general changes, toxicity, and fatality rate. The side effects and dying animals were not observed by NY945. Seven or 14 days after observation of general behavioral state, the body weight of the animals pretreated with NY945 was not different from that of the control group.

When NY945 (100-500µg/ml, 1mg/kg) was treated to the isolated ileum, trachea, lung or right atrium tissues, there were no effects on the contractility of each tissue, the heart rate or the contractile force of the right atrium. The contractility and dilatation of each isolated tissue caused by various agonists (epinephrine, isoproterenol, acetylcholine, histamine, or barium chloride) were also not affected by NY945.

### 3. Sub-acute toxicity test

Healthy guinea pigs or rats were selected after observation of the general changes. Experimental groups were divided into the PBS-treated group (control), low-dose NY945 (40µg/kg)-treated group, moderate-dose (200µg/kg)-treated group, and high-dose (1mg/kg)-treated group. Each group (N = 10 animals/group) underwent repeated administration with intradermal injection one time per week for one month. During injection, we examined the symptoms, amounts of diet and water uptake, body weights, eye test, urine tests (volume, color, glucose, pH, etc.), and hematological tests (general hematological test and biochemical test). After the course of injection was finished, all these tests were repeated. None of the test results showed any change with each dose of NY945 compared to the control group, and there were no differences between guinea pigs and rats.

Each organ (liver, spleen, pancreas, kidneys, adrenal glands, ovaries, heart, lung, thyroid gland, and brain pituitary gland) isolated from the sacrificed animal was weighed and underwent histological testing. None of the NY945-treated groups showed any pathophysiological lesion, compared to the control group.

## 4. Immune toxicity test
### 1) Anaphylactic shock reaction

In order to examine the anaphylactic shock reaction for NY945, NY945 (500μg/kg, 1mg/kg, 3mg/kg, 5mg/kg) alone was administrated with i.p. injection in guinea pigs, compared to the positive control group (OVA-induced group). OVA-induced mice induced anaphylactic shock, but NY945 alone did not (Table 3).

NY945 (3mg/kg) protected the OVA-induced anaphylactic shock reaction in a preliminary study. Therefore, the experimental groups were divided into 7 groups: saline alone-treated, OVA-adsorbed alum alone-treated (positive control, OVA group), NY945 (1mg/kg) plus OVA, NY945 (3mg/kg) plus OVA, NY945 (5mg/kg) plus OVA, NY945 (1mg/kg) plus CFA plus OVA, and NY945 (3mg/kg) plus CFA plus OVA. Anaphylactic shock was induced under the standard condition of "Toxicity Test" as presented by the Korean Food & Drug Administration (KFDA) (Number 96-8). All NY945-treated OVA groups were completely protected against anaphylactic shock. CFA-sensitized groups were not completely protected, but showed mild anaphylactic symptom (Table 3).

### 2) Passive cutaneou anaphylaxis

Serum was collected from the 7 guinea pig groups described in the previous paragraph of section 1 above. These sera containing anti-OVA antibodies were used for the passive cutaneous anaphylaxis test. The OVA alone-treated group (positive control) showed strong antibody (IgE) titer using serum dilution test (titer, 2,500 ~ 3,200), but none of the NY945 groups showed anti-OVA antibody titer (titer, 500 ~ 1,000).

**Table 3.** Experimental design of anaphylactic shock and symptoms of guinea pig after OVA-Al(OH)$_3$ conjugates or NY945 challenge[a].

| roup[b] | Pre-treatment | Sensitizing antigen | Challenging antigen | Severing of anaphylaxis | | | | |
|---|---|---|---|---|---|---|---|---|
| | | | | - | ± | + | ++ | +++ |
| 1 | Saline | Saline | Saline | 6 | -[c] | - | - | - |
| 2 | Saline | OVA-Al(OH)$_3$ (100μg /0.2ml) | OVA-Al(OH)$_3$ (100μg /0.2ml) | - | - | - | 1 | 5 |
| 3 | NY945 1mg/kg | OVA-Al(OH)$_3$ (100μg /0.2ml) | OVA-Al(OH)$_3$ (100μg /0.2ml) | 6 | - | - | - | - |
| 4 | NY945 3mg/kg | OVA-Al(OH)$_3$ (100μg /0.2ml) | OVA-Al(OH)$_3$ (100μg /0.2ml) | 6 | - | - | - | - |
| 5 | NY945 5mg/kg | OVA-Al(OH)$_3$ (100μg /0.2ml) | OVA-Al(OH)$_3$ (100μg /0.2ml) | 6 | - | - | - | - |

| | | | | - | ± | + | ++ | +++ |
|---|---|---|---|---|---|---|---|---|
| 6 | NY945 1mg/kg[e] | OVA-Al(OH)$_3$ +CFA | OVA-Al(OH)$_3$ (100µg /0.2ml) | 4 | 2 | - | - | - |
| 7 | NY945 5mg/kg | OVA-Al(OH)$_3$ +CFA | OVA-Al(OH)$_3$ (100µg /0.2ml) | 5 | 1 | - | - | - |

[a]Each group was sensitized with s.c. injection following experimental design for 4wks.Each antigen was challenged by i.v. injection 3wks after final injection, and observed the degree of symptoms. -, asymptomatic; ±, mild; +, moderate; ++, severe; +++, death.
[b]Number of animal in each group was 6 guinea pigs.
[c]Zero number of animal in each group
[d]OVA-Al(OH)$_3$ conjugate contains 100µg ovalbumin (OVA) and 125µg Al(OH)$_3$.
[e]NY945-CFA (Complete Freund's Adjuvant) conjugate.

## 5. Skin irritation test

According the standard of toxicity test and drug etc. in section number 96-8 (1999. 4. 16.) of KFDA, NY945 (3, 5mg/kg) was applied to mouse skin one time/week for 4 weeks and covered with gauge after shaving. Various targets were then examined.

### 1) Proliferation of splenocytes

After application of NY945 (3mg/kg or 5mg/kg) alone, splenocytes were isolated and examined to determine whether the cells had proliferated or not. NY945 alone did not affect the proliferation of splenocytes. However, the OVA alone- and OVA adsorbed alum-treated group experienced increased proliferation of splenocytes, and cell proliferation for NY945 was increased more than that of the OVA group (Table 4).

### 2) Production of cytokines

Application of NY945 alone did not affect the production of INF-$\gamma$ and IL-4. However, OVA alone and OVA plus alum groups remarkably increased the production of INF-$\gamma$, whereas IL-4 did not. NY945 pretreatment 1hr before OVA challenge reduced INF-$\gamma$ production. NY945 alone application did not affect the production of IgE antibody, but did remarkably inhibit the elevated IgE caused by OVA challenge (Tables 4 and 5).

### 3) Skin histology

NY945 alone did not show infiltration of eosinophils into the skin mucosa, but it did inhibit the increased eosinophil infiltration caused by OVA challenge (data not shown).

## 6. Chronic toxicity

Chronic toxicity (repeated injection for 3 months) was not tested due to difficulty in purifying a large amount of NY945. In the future, it will be performed after production of recombinant alprogen.

**Table 4.**   Effect of NY945 on the cytokines, serum IgE antibody level and proliferation in the mouse epicutaneous immunization model.[a]

| OVA(ug/ml) | IL-4 (pg/ml) | IFN-$\gamma$ (pg/ml) | IgE Ab (O.D) | Proliferation ( x $10^3$ cpm) | | | |
| --- | --- | --- | --- | --- | --- | --- | --- |
| | | | | 0 | 0.1 ug | 1.0 ug | 10.0 ug |
| Saline | -[b] | 50.3 ± 9.70 | 0.08 ± 0.001 | 1.3 ± 0.19 | 1.3 ± 0.18 | 1.5 ±0.27 | 1.2 ± 0.10 |
| OVA + Alum | 363.8 ± 69.09 | 818.0 ± 14.21 | 3.32 ± 0.005 | 1.3 ± 0.37 | 1.4 ± 0.31 | 1.5 ± 0.18 | 2.0 ± 0.17 |
| OVA alone | 16.2 ± 2.71 | 435.8 ± 84.93 | 0.92 ± 0.110 | 1.0 ± 0.12 | 0.9 ± 0.08 | 1.0 ± 0.14 | 1.8 ± 0.10 |
| NY945 + OVA Pre 5mg/kg | 14.8 ± 1.69 | 536.1 ± 188.82 | 0.11 ± 0.012 | 1.5 ± 0.32 | 1.9 ± 0.53 | 1.9 ± 0.48 | 1.1 ± 0.04 |
| 5mg/kg (1hr) | 15.7 ± 3.94 | 65.8 ± 11.68 | 0.57 ± 11.680 | 1.1 ± 0.67 | 0.9 ± 0.24 | 1.0 ± 0.29 | 1.5 ± 0.08 |
| 3mg/kg (1hr) | -[b] | 184.4 ± 31.02 | 0.58 ± 31.020 | 0.9 ± 0.23 | 0.9 ± 0.21 | 1.0 ± 0.18 | 1.6 ± 0.14 |
| NY945 alone5mg/kg | -[b] | 338.2 ± 83.26 | -[b] | 1.2 ± 0.35 | 1.0 ± 0.30 | 1.2 ± 0.07 | 0.9 ± 0.15 |
| 3mg/kg | -[b] | 130.7 ± 25.25 | -[b] | 1.4 ± 0.24 | 1.8 ± 0.62 | 2.1 ± 0.48 | 1.9 ± 0.51 |

[a]Mice (BALB/c) were immunized by the commonly used patch methods. Mice were anesthetized with avertin before being shaved on the back by an electric blade. Ovalbumin (OVA, 100µg/100µl) or NY945 (3mg or 5mg/kg) was first applied to patches sized 1cm$^2$, which were then applied to the shaved skin, fixed and secured with an elastic bandage. For each course of immunization, patches were left on from day 0 (Ed- confirm this addition) to 4, and patches prepared freshly were used from day 4 to day 7. Each course of immunization was repeated 5 times at 2wk intervals. NY945 was pretreated intraperitoneally 1wk before immunization or 1hr before the mice were sacrificed. Cytokines and IgE were measured by ELISA method from isolated splenocytes (4 x $10^6$ cells for cytokines, 1 x $10^6$ cells for proliferation).
[b] No detection. **, P<0.01; ***, P<0.001 by comparison with saline. +, P<0.05; ++, P<0.01 by comparison with OVA alone.

**Table 5.** Effect of NY945 on the symptom score, serum IgG and IgE antibody levels in the guinea pig asthma model[a].

| | Hyperpnea & Dyspnea | | | | | Serum total IgG titer[b] | Serum total IgE titer[c] |
|---|---|---|---|---|---|---|---|
| | - | ± | + | ++ | +++ | | |
| Saline | 6 | - | - | - | - | - | - |
| OVA alone | - | - | - | 4 | 2 | 65.6 ± 19.0 | - |
| Ox-Ascaris ˙ | | | | | | - | 2,240 ± 392 |
| NY945 | | | | | | | |
| 3mg/kg | 2 | 3 | 1 | | | 28.8 ± 3.20 | 960 ± 271**(57.1)[d] |
| 5mg/kg | 5 | 1 | | | | 41.6 ± 9.60 | 680 ± 250**(69.6) |

[a]Guinea pigs were sensitized and sprayed with OVA (20µg containing 4.5mg alum) with the same methods described above. Allergic symptom scores (sneezing and rubbing nose) were counted for 30min after OVA challenge.

[b]Serum total IgG titer was determined by PCA method.

[c]Guinea pigs received i.p. injections of 250mg/kg of cyclophosphamide 2days before primary i.p. immunization with 1µg of conjugate (Ox-Asc) adsorbed to 1mg of $Al(OH)_3$ (Alum) in 1ml of saline. Every month thereafter for 5 months, a similar dose of antigen in alum was administrated i.p. At the end of 5 months the animals was exsanguinated. The titer of sera obtained from the bleeding was measured by PCA method.

[d]Parentheses are the decreasing percentage evoked by administration of NY945.

**, $P<0.01$ by comparison with OVA challenge.

### 7.Genetic toxicity

Professor Jong Koo Kang at Chungbuk University Veterinary School (GLP organization) investigated the reverse mutation study, chromosome aberration study and micronucleus test.

### 1) Reverse mutation study using bacteria

NY945 (56.3, 112.5, 225, 450, 900μg/plate) did not affect either the metabolizing system or proliferation of the colony number using Salmonella typhimurium and Escherichia coli without S9 mix. The same result was found in a dose-dependent manner. In the positive control using sodium azid (SA), 9-aminoacridine (9-AA), 2-aminoanthracene (2-AA), or 2-aminofluorene (AF-A), the number of reverse mutation colonies was increased 2-fold, compared to the negative control. These data suggest that NY945 is not able to induce reverse mutation.

### 2) Chromosome aberration study using Chinese-hamster ling (CHL)-cultured cells

Twenty-four hours after NY945 (300, 900μg) was added to cultured cells, chromosome aberration, and aberration of structural or numerical frequency were not affected independent of metabolizing activity. Therefore, these data suggest that NY945 does not induce either structural or numeral aberrations in chromo-somes independent of metabolizing activity.

### 3) Micronucleus test using bone marrow cells in male ICR mice

NY945 (100, 300, 900μg) did not significantly induce bleeding of micro-nucleated polychromatic erythrocyte (MNPCE) among PCE. Compared to the negative control group, the ratio of PCE/total RBC was increased by 30%, and cellular toxicity by NY945, such as inhibition of hematopoietic function etc., was also not observed. Therefore, these data suggest that NY945 does not induce formation of micronucleated PCE in mouse bone marrow cells.


## 4.3.2.6. Test for various properties of matter after preparation of nebulizer

### 1. Preparation of nebulizer

We prepared 2mg/10ml NY945 nebulizer (10 times for each dose). The optimal pH of NY945 nebulizer was 8.6.

### 2. Measurement of solubility

NY945 was easily dissolved in ethylalcohol (organic solvent) or $H_2O$. That is, other solvents were not required because NY945 was dissolved very well by water, to a possible dilution limit of 100,000 times in water.

### 3. Analysis by HPLC

The results of analysis by HPLC gradient system (μ Bondapak C18 column) after water dilution of NY945 yielded a linear calibration curve.

### 4. Interaction by composition of additives

#### 1) Possibility of container composition in the absence of preservative

Safekeeping test for NY945 nebulizer confirmed the possibility of use without other preservatives.

#### 2) Composition possibility of NT945 container

NY945 original solution, polyethylene nebulizer, and 2mg/10ml NY945 nebulizer were incubated at room temperature, after which they were confirmed to be harmless by HPLC analysis.

#### 3) Heating of NY945 nebulizer preparation

NY945 was heated to 150°C and analyzed using differential scanning colorimetry (DSC). We confirmed that the preparation of the container containing NY945 or nebulizer was harmless.

### 5. Safekeeping of nebulizer during short or long period

#### 1) Safekeeping for a short period : NY945 nebulizer was harmless at room temperature or after 30-day incubation at 40°C.

#### 2) Safekeeping for a long period

(1) Preparation of anti-alprogen antibody: First, we prepared anti-alprogen antibody via immunization of rabbit as a safekeeping test. Titers of anti-alprogen antibody in rabbit sera were measured by PCA method. These sera were used to measure the safety of NY945 nebulizer for a long period of incubation.

(2) NY945 was incubated at room temperature or 40°C for 1, 2, 3, 4, 5, and 10 months, and then safekeeping test was performed by ELISA method using anti-alprogen antibody. The results showed that incubation safekeeping of NY945 at room temperature remained safe for 3 months, whereas incubation safekeeping at 40°C was safe for only 2 months.

### 6. Selection of first packing unit

The optimal packing unit was a nebulizer because it has to be reached from the nasal area or throat to alveoli. Particle size was measured by an international public organization, Saint-Gobain Company, was an average of 46.80μm. This particle size is sufficient to be reached by alveoli.

### 7. Animal model test of nebulizer preparation

An asthma model was prepared in mice and guinea pigs by i.p. injection of OVA and nebulization of OVA for 49 days, as described in "previous method". In the previous animal model method, NY945 (3mg/kg or 5mg/kg) was pretreated by i.p. injection 1hr before OVA challenge or 1 week before OVA sensitization, whereas in the nebulizer preparation, NY945 was nebulized in the nasal area or throat. The results were as follows.

### 1) Mouse asthma model

(1) NY945 nebulizer (3mg/kg or 5mg/kg) reduced the increased OVA-induced asthma symptom score by 52.5% and 64.9%, respectively, and also reduced the serum IgE level in a dose-dependent manner.

(2) NY945 nebulizer (5mg/kg) reduced relative TAR.

(3) NY945 nebulizer (3mg/kg or 5mg/kg) remarkably reduced the infiltration of eosinophils and mast cells into nasal septic mucosa or lung tissues.

### 2) Guinea pig lung asthma model

(1) NY945 nebulizer (3mg/kg or 5mg/kg) reduced the increased OVA-induced asthma symptom score and TAR, and encouraged recovery from hyperpnea and dyspnea. NY945 nebulizer also reduced serum IgE level by 54% and 75%, respectively.

(2) NY945 (3mg/kg or 5mg/kg) remarkably inhibited the infiltration of eosinophils and mast cells into nasal septic mucosa or lung tissues.

## 4.3.2.7. Other effects of NY945 of alprogen
### 1. Anti-gastric ulcer effect of alprogen

Alprogen (5.0µg) weakly inhibited the increased $LTD_4$ and $LTB_4$ produced from *Helicobacter pylori* (*H. pylori*)-stimulated, or A23187-stimulated neutrophils or the gastric mucosa cell line (Kato □ cells).

### 2. Anti-gastric ulcer effect of NY945

### 1) Inhibitory effect and mechanism of mediator release

NY945 (200µg concentration) weakly inhibited the increased $LTD_4$ and $LTB_4$ produced from *H. pylori*-stimulated or A23187-stimulated neutrophils or Kato β cells, and also weakly inhibited the increased intracellular $Ca^{2+}$ level during stimulation of both cells. The degree of inhibition showed a similar result to the low concentration of NY945 (50µg).

### 2) Anti-oxidative effect of NY945

NY945 (5, 50, 100, and 200µg) did not affect NADPH oxidase or MPO (myeloperoxidase) activities, which are known as targets of reactive oxygen species (ROS), in the KATO β cells or neutrophils stimulated by *H. pylori* or A23187.

These data suggest that NY945 or alprogen has a weak anti-gastric ulcer effect.

### 3. Anti-oxidative effect of alprogen

When the sensitized guinea pig lung mast cells were challenged with 10µg/ml OVA, ROS production peaked. The produced ROS was inhibited by alprogen by 40%. This result was similar to the effect of SOD, a previously known antioxidant.

The concentration of antigen (OVA) used was 10 times higher than that used for mediator release.

### 4. Cellular protein analysis related to the allergic diseases by NY945

HMC-1 cells were sensitized and challenged with anti-IgE antibody and hIgE, respectively. Proteins were separated by two-dimensional electrophoresis, using immobilized pH gradients of 3~10. After that, the proteins were stained with Coomassie or silver stain and the stained gels were analyzed with software (Image master) and MALDI-TOF. The number of protein expression increased when the HMC-1 cell activation was approximately 100 spots by Coomassie blue staining, and approximately 250 spots by silver staining. Alprogen (5μg) pretreatment reduced the expressions of 6 proteins among 100 spots and of 8 proteins among 250 spots, respectively. The proteins procured by Coomassie blue staining were heat shock protein 60 (HSP 60), ATP synthase β chain (mitochondrial precursor), tropomycin 4, cytoskeletal tropomycin, calbindin 2, and tumor protein (translationally-controlled 1), and those by silver staining in addition were calbindin and prohibitin

HSP 60 is known to release various cytokines in cells, and these cytokines activate macrophages and endothelial cells to elevate the expression of adhesion molecules, which induce the infiltration of inflammatory cells such as eosinophils, mast cells, and neutrophils, etc. into inflammatory sites. It has been known that calbindin 2 or tumor protein is the binding protein, and prohibitin promotes maturation of B cells. These data suggest that all proteins found from proteomics analyses were related to allergic diseases, and that they may also have other functions. Many of the spots observed by silver staining were very small. Therefore, it was very difficult to confirm the proteins.

### 5. Effects of NY945 on rheumatoid arthritis (RA)

In order to examine the effect of NY945 on RA, we prepared a collagen (Type iv)-induced RA model in Lewis rats. We examined the RA factors as follows.

### 1) Edema formation of hind paw

Edema size (redness/swelling) of murine collagen-induced hind paw was remarkably increased. NY945 (5mg/kg) pretreatment at 1week or 1hr before RA induction with collagen inhibited edema size and redness.

### 2) Inflammatory or immune cells in blood

In the RA model, body weight was not decreased, but the number of eosinophils and lymphocytes was increased. NY945 (5mg/kg) decreased both cells.

### 3) Production of inflammatory cytokines or antibody

Amounts of TNF-α or IL-1β containing serum were similar to the level in normal rat serum. NY945 (5mg/kg) did not affect the production of TNF-α or IL-1β. However, NY945 inhibited the level of anti-collagen antibody increased in collagen-induced RA serum.

### 4.3.3. Conclusion

This work was supported by the research fund of KISTEP and Nam Yang Company for 7 years. Single glycoprotein, alprogen, and the pre-step fraction of alprogen, NY945, were separated from *Aloe vera* whole-leaf. These components have an anti-allergic effect on mediators such as histamine, leukotriene ($LTD_4$), and cytokines (TNF-$\alpha$, IL-8, IL-4 etc) and the mechanisms of this effect were studied. It was also demonstrated that NY945 was effective in OVA-induced mouse or guinea pig allergic rhinitis or asthma model (in vivo). Toxicity tests such as acute, sub-acute, 1 month chronic, immune, genetic, and cutaneous tests were performed, and NY945 nebulizer was prepared. Other effects of NY945 or alprogen on the anti-gastric ulcer effect, anti-oxidative effect, and anti-rheumatoid arthritis were studied. Furthermore, as shown in Fig. 5, we observed that all enzymes, messengers, mediators, and cytokines of rectangular and elliptic shapes were inhibited by NY945 or alprogen.

It can be inferred that alprogen binds the membrane of mast cells, or binds the membrane-bound IgE antibody, after which the membrane-bound alprogen interferes with the mobility of receptors for antigen/antibody cross-linking or the IgE antibody-bound alprogen interferes with the antigen to bind antibody (IgE). Therefore, alprogen will inhibit the cascade of multi-signal pathways in mast cells activated with antigen/antibody reactions (Fig. 5)

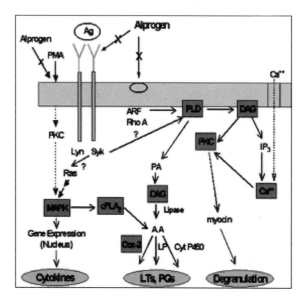

**Fig. 5.** Possible action site of Alprogen and pathway of signal transduction in mast cell activation. (See Plate 14.)

# References

1. Abbas AK, Lichtman AH, Cellular and molecular Immunology. 5th 3rd, Elservier Science (SAUNDERS), Philadelphia, pp. 432-452, 2003
2. Altrichter J, Guse AH, Resch K, Brock J, Daeron M, Huckel C. Phosphatidy-linositol hydrolysis and an increase in $Ca^{++}$ concentration in the signal-transduction process triggered by murine FcγRIII are not required for protein kinase C translocation. Eur J Biochem 228: 587-595, 1995
3. Choi WS Kim YM, Combs C, Frohman MA, Beaven MA. Phospholiphase D1 and D2 regulate different phases of exocytosis in mast cells. J Immunol 168:5683-5689, 2002
4. Davis RH, Didonato JJ, Hartman GN, Haas RC, Anti-inflammatory and wound healing activity of a growth substance in Aloe vera. JAPMA, 84(2): 77-81, 1994
5. Fonteh AN, Atsumi G, LaPorte T, Chilton FH. Secretory phospholipase $A_2$ receptor-mediated activation of cytosolic phospholipase $A_2$ in murine-derived mast cells. J Immunol 165:2773-2782, 2000
6. Goldsby RA, Kindt TJ, Osborne BA, Kuby J, Immunology. 5th eds., W.H. Freeman and Comp, New York, pp. 361-378, 2003
7. Gould HJ, Gilfilan Am, Reischi IG, Beavil R. IgE structure, receptors and signaling. In Allergy, eds. Holgate ST, Church MK, Lichtenstein LM 2nd eds, Mosby, London, pp. 243-264, 2001
8. Kettner A, Kumar L, Anton IM, Sasahara Y, De La Fuenta M, Pivniouk VI Falet H, Hartwig JH, Geha RS. WIP regulates signaling via the high affinity receptor for immunoglobulin E in mast cells. J Exp Med 199:357-368, 2004
9. Karaca K, Sharma JM, Nordgren R, Nitric oxide production by chicken macrophages activated by acemannan, a complex carbohydrate extract from Aloe vera. Int'l J Immunopharmacol, 17(3):183-188, 1995
10. Kim JY, Lee KH, Lee BK, Ro JY. Peroxynitrite modulates release of inflammatory mediators from guinea pig lung mast cells activated by antigen-antibody reaction. Int Arch Allergy Immunol 137:104-114, 2005
11. Kim JY, Ro JY. Signal pathway of cytokines produced by reactive oxygen species generated from phorbol myristiate acetate-stimulated HMC-1 cells. Scan J Immunol 62:25-35, 2005
12. Reuben P Siragainan, Biochemical events in basophil/mast cell activation and mediator release. In Allen P Kaplan ed., Allergy, 2nd eds., New York, W.B. Saunders Comp, pp. 99-131 (1997)
13. Ro JY, Lee BC, Kim JY, Chung MH, Lee SK, Kim KH, Park YI, The inhibitory mechanism of Aloe single component (Alprogen) on the mediator release in the guinea pig lung mast cells activated with specific antigen-antibody reactions. J Pharmacol Exp Ther, 292:114-121, 2000
14. Williams MS, Burk M, Loprinzi CL, Hill M, Schomberg PJ, Nearhood IC, OFallon JR, Laurie JA, Shanahan TG, Moore RL, Urias RE, Kuske RR, Engel RE, Eggleston WD, Phase III double-blind evaluation of an Aloe vera gel as a prophylactic agent for radiation-induced skin toxicity. Int'l J Rad Oncol Bio Phys, 36(2): 345-349, 1996

15. Zhang L, Tizard IR, Activation of a mouse macrophage cell line by aceman-
    nan: The major carbohydrate fraction from Aloe vera gel. Immunopharmacol,
    35(2): 119-128, 1996

## 4.4. Skin whitening effect

### Park, Young In, Ph.D.

School of Life Sciences and Biotechnology, Korea University

Humankind's economic activities are damaging the earth's environment at an increasing rate. Air pollution is becoming a serious issue because of extensive use of fossil fuels, reduction of green zones due to urban development, and the extensive use of the now-banned coolant gases such as freon gas. These aspects resulted in the stratospheric ozone depletion, thereby increasing the level of UV irradiation on the ground which has potentially harmful effects to the human body. UV causes many mutations among life forms on the Earth. Some of the harmful effects of UV irradiation on the human body include eye cataracts and various diseases such as skin cancer, facial pigmented spots and freckles. In addition, as the demand for higher living standards increases, people are more prone to be engaged in outdoor activities which increase their exposure to UV irradiation. Although hyperpigmentation of the skin including discoloration on the face is not caused only by UV irradiation, this is becoming a serious health and beauty problem as more and more people are exposed to UV and its effects on the human body.

At present, two kinds of method are, in general, applied to reduce hyperpigmentation of the skin in cosmetics. The first approach is the use of UV blockers, which prevents UV rays from reaching the skin by applying cream-like UV blocking substances such as sun screen. The second way is the use of enzyme inhibitors which block the biosynthesis of melanin in human skin melanocytes. In other words, the skin whitening effect is achieved by inhibiting the activity of tyrosinase, an enzyme responsible for the synthesis of melanin.

Specks or freckles on the skin caused by the over-production and localization of melanin pigment are due to the activation of tyrosinase that exists in melanosome of the melanocytes on the dermal skin (Friedman et al. 1987). Melanin is synthesized from tyrosine by monophenol monooxygenase (tyrosinase) [EC 1.14.18.1] by way of derivatives of quinone or indolequinone (Pawelek 1976; Hearing 1987; Fig. 1). Namely, tyrosinase is a copper-containing enzyme which catalyzes the ortho-hydroxylation of monophenols and the oxidation of o-diphenols to o-quinones. This enzyme thus catalyzes the first two steps in the synthesis of melanin by converting tyrosine into DOPA (3,4-dihydroxyphenylalanine). The rest of the pathway is known to proceed almost automatically for the synthesis of melanin. Melanin pigments are, in fact, present as a polymerized form of heterogenous complexes which are composed of many subunits originating from various precursor molecules. So far, the exact composition and structure of melanin has not been elucidated but eumelanin typically expresses a black color. However, eumelanin can be converted to pheomelanin, which is a yellow or brown color, by copolymerization of sulfur- containing amino acid, cysteine, and DOPA (Prota et al. 1970). However, the exact cellular mechanism which controls the composition of various melanin pigments has not yet been elucidated. Tyrosinases have been found in various organisms such as mammals, lower eukaryotes, plants and fungi, but their activities are presumably not influenced by substrate specificity or cofactors.

As explained above, tyrosinase serves as a key enzyme (as a mixed function oxidase) that catalyzes the first two steps in melanin biosynthetic pathway, i.e. hydroxylation of L-tyrosine to L-DOPA and of L-DOPA to dopaquinone. In addition, much research has been conducted on melanin synthesis through tyrosinase and the growth of melanocytes. As a result, activation of tyrosinase through stimulation of cell growth by melanocyte stimulating hormones (MSH) and adenylate cyclase activation within cells have been reported (Bryan et al. 1987; Pawelek 1977). Research on the development of depigmenting (or skin whitening) agents is actively underway in terms of inhibiting melanocyte cell division or inhibiting the activation of tyrosinase enzyme. Recently, tyrosinase inhibitors extracted from various microorganisms or plants have been reported and are either being used or being developed for the use as depigmenting agents. Those agents currently being used in cosmetics for skin whitening purpose are hydroquinones, catechols, vitamin C, arbutin, which is a hydroquinone glycoside isolated from Uva ursi leaves, and kojic acid, which is extracted from the fermented soybean by kojic mold. Especially, arbutin is known to inhibit tyro-sinase competitively whereas kojic acid inhibits tyrosinase activity by forming chelation with copper present in tyrosinase (Maeda and Fukuda 1996).

However, it is difficult to use kojic acid in cosmetics because of its bad smell, dark color and various side effects such as skin irritation.

However, the increasing usage of skin whitening agents for cosmetics is mostly based on the inhibition of tyrosinase which acts on the rate limiting steps in the synthesis of melanin pigments. Interestingly, the aloe extract being used as a basic ingredient in cosmetics has been reported to contain mushroom tyrosinase inhibitor (Yagi et al. 1986). Based on this report, we attempted to isolate tyro-sinase inhibitor from Aloe vera.

In order to isolate the skin whitening component from aloe, the methodology of assay for tyrosinase enzyme activity was established. There are two traditional methods of assay: one is a modified method of the charcoal adsorption used by Pomerantz (1966), known as the 'Pomerantz Method', and the other is the 'Dopachrome Method' used by Mason (1948).

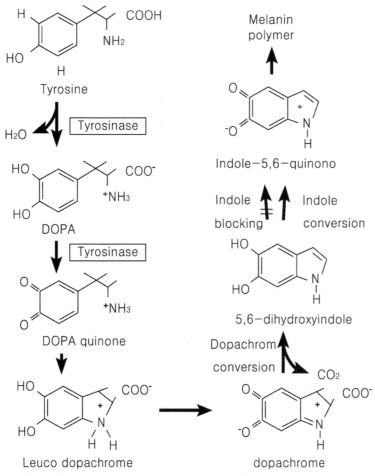

**Fig. 1.** Biosynthetic pathway of melanin DOPA :3,4-dihydroxindolephenylalanine, Dopaquinone : 4-(2-carboxyl-2-aminoethyl)-1,2-benzoquinone
Dopachrome : 2-carboxyl-2,3-dihydroindole-5,6-quinone

Experiments were conducted through these procedures and an optimum assay condition was established. Namely, using arbutin and kojic acid as positive controls, the respective inhibiting activity for tyrosinase was assayed. As a result, inhibition of tyrosinase activity was detected in both materials by the 'Pomerantz Method'. However, inhibition of tyrosinase activity by arbutin was not observed through the 'Dopachrome Method'. Therefore, the modified 'Pomerantz Method' was chosen to screen fractions extracted from the inhibition of tyrosinase activity. Aloe dry powder (so called, 200:1 gel powder) was extracted with 70% ethanol, which showed inhibition activity against tyrosinase. This ethanol extract was then further fractionated with various solutions such as hexane, chloroform, ethylacetate and butanol. The soluble fraction of butanol extraction showed the activity, and was therefore applied to silica gel column chromatography with ethylacetate/methanol/$H_2O$ (5:1:0.1) as a developing solution (Fig. 2). Finally, the fraction 1-5-3 was recrystallized with a mixed solution of ethanol/ethylacetate/methanol. The resulting compound was identified as aloesin by mass and NMR and was designated as 'aloewhite'.

To develop aloewhite as a skin whitening agent, various experiments such as assessment of its toxicity and stability, in vitro biochemical studies, in vivo animal tests and human trials have been conducted. It has been determined that the level of inhibition of tyrosinase activity was proportional to the concentration of aloewhite in a dose dependent manner: the tyrosinase enzyme activity decreased with increasing aloewhite concentration (Fig. 3, Table 1). Especially, at 0.5 mM of aloewhite concentration, a suppression rate of 68% was observed, and at 5 mM the suppression rate was over 90%. In addition, to identify aloewhite's exact inhibition mechanism of tyrosinase enzyme activity, an enzyme kinetic study was conducted. As a result, aloewhite suppressed tyrosinase activity through a noncompetitive manner, and its Ki value was determined to be at $5 \times 10^{-4}$ M (Fig. 4).

Based on reports showing that both arbutin and kojic acid inhibit tyrosinase enzyme activity competitively, it was necessary to determine whether any combination of these two substances would increase the inhibition of tyrosinase activity synergistically. The inhibition of enzyme activities by the treatment with 0.1 mM of aloewhite, arbutin or kojic acid alone was 49%, 62.4% or 64.1%, respectively. A combination of 0.03 mM aloewhite and 0.03 mM arbutin inhibited the tyrosinase activity by 89.5%, whereas a combination of 0.03 mM aloewhite and 0.03 mM kojic acid inhibited the enzyme activity by 77.6% (Fig. 3, Table 2). These results demonstrated that the combination of arbutin or kojic acid with aloewhite can synergistically inhibit tyrosinase activity and hence, that the effective dosage can be significantly reduced for the same inhibitory activity against tyrosinase (Jin et al. 1999).

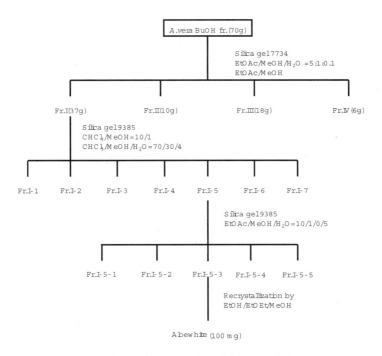

**Fig. 2.** Preparation scheme of aloewhite from butanol fraction of Aloe vera

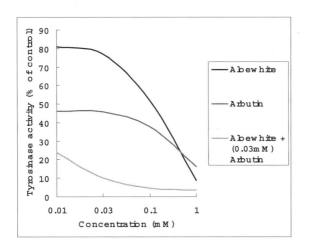

**Fig. 3.** Dose-dependent inhibition of aloewhite and arbutin on the tyrosinase activity    Each data is the mean of triplicate determinations ± SE The tyrosinase activity was determined by Pomerantz method. (See Plate 14.)

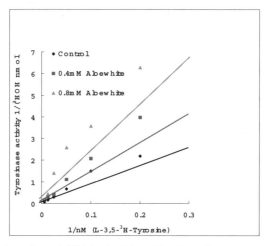

**Fig. 4.** Kinetics of tyrosinase inhibition by Aloewhite. (See Plate 15.)

**Table 1.** Kinetics of tyrosinase inhibition by Aloewhite at 0.4 and 0.8mM.

| 1/L-3,5-3H-tyrosine(nM) | 1/3HOH(nmol) | | |
|---|---|---|---|
| | Control | 0.4mM Aloewhite | 0.8mM Aloewhite |
| $2.0 \times 10^{-1}$ | 2.20 | 3.97 | 6.28 |
| $1.0 \times 10^{-1}$ | 1.50 | 2.08 | 3.58 |
| $5.0 \times 10^{-2}$ | 0.67 | 1.10 | 2.59 |
| $2.5 \times 10^{-2}$ | 0.31 | 0.42 | 1.42 |
| $1.3 \times 10^{-2}$ | 0.17 | 0.30 | 0.43 |
| $6.3 \times 10^{-3}$ | 0.08 | 0.16 | 0.18 |

* The tyrosinase activity was determined by Pomerantz method

**Table 2.** Dose-dependent inhibition of Aloewhite and Arbutin on the tyrosinase activity

| Conc.(mM) | % of control | | |
|---|---|---|---|
| | Aloewhite | Arbutin | Aloewhite + (0.03mM) Arbutin |
| control | $100 \pm 7.9$ | $100 \pm 7.9$ | $45.8 \pm 0.9$ |
| 0.01 | $81.0 \pm 1.0$ | $46.1 \pm 1.9$ | $23.6 \pm 3.8$ |
| 0.03 | $77.0 \pm 1.8$ | $45.8 \pm 0.9$ | $10.2 \pm 0.7$ |
| 0.1 | $51.5 \pm 1.4$ | $37.6 \pm 0.3$ | $4.5 \pm 0.4$ |
| 1 | $8.8 \pm 1.0$ | $16.1 \pm 1.0$ | $3.6 \pm 0.7$ |

The effects of these substances on skin whitening were examined after UV ir-radiation on human skin. The test was conducted by irradiating UV-B (280-315 nm) at 210mJ (2.34 mJ/sec × 90 sec) over a piece of corrugated cardboard with four 12mm circular holes on the inside of the upper part of an arm. After the first course of 24-hour UV irradiation ointments containing either arbutin, aloewhite or a combination of both were applied 4 times daily (in the morning, afternoon, eve-ning, and before sleep) for 14 days in order to compare the degree of depigmenta-tion in the 4 spots. The ointment was composed of a water-soluble base containing substances at the concentration of 0.2 mg/ml. From the results, the combination of aloewhite and arbutin showed the best skin whitening effect (Fig. 5).

In order to be used as an ingredient for cosmetics, aloewhite must pass the toxi-city and safety tests to confirm that it is basically harmless to the human body. To achieve this result, aloewhite must undergo various tests such as acute toxicity test by oral administration, acute subcutaneous test, acute dermal irritation test, ocular irritation test, skin sensitization test, etc., according to the general toxicity test guidelines established by the standard operation procedure of the KGLP (Korean Good Laboratory Procedure 1992). The tests were successfully conducted to meet all these regulations and standards, i.e., Notice 94-3 of the National Institute of Safety Research (currently renamed as the National Institute of Toxicological Re-search). Because no significant side-effects from use were found, aloewhite acquired a certificate as an acceptable skin whitening component for cosmetic use from KFDA (Korea Food and Drug Administration).

As a result " RAMENT Aloe Whitening" has been developed to contain aloe-white as a skin whitening substance. Currently, various other depigmenting sub-stances such as vitamin C and placenta extract are being tested. However, in general, they are either less effective or their toxic levels render them incompati-ble with current test standards. Especially in the case of kojic acid, because it is extracted mostly from soy sauce, it emits a very distinctive odor and dark color which are not proper for cosmetic use. In addition, kojic acid is suspected to be carcinogenic. Compared to kojic acid, aloewhite shows high quality and compati-bility for use in cosmetics.

As economic growth continuously increases the elderly population, concern about skin health is rising among the aged. Moreover, the increasing participation in outdoor activities may increase the exposure to UV irradiation on human skin, thereby increasing the frequency of speckles and freckles. In addition, the aware-ness of the harmful effects of UV light is rising among the general public. This has contributed to the ever increasing demand and consumption of cosmetics and a change in the standard food consumption patterns. From the simple solution in the 1980s of merely preventing harmful UV irradiation from reaching the skin, the demand for an all-year usable skin whitening cosmetics that contain both func-tions of UV protection and skin whitening is increasing these days.

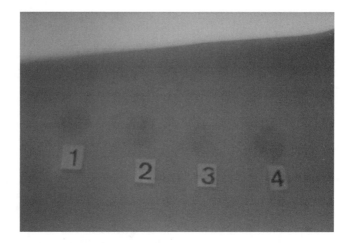

**Fig. 5.** *In vivo* assay for the whitening effect on human skin. (See Plate 15.)
1. negative control (++++)
2. Aloewhite administered (++)
3. Aloewhite + Arbutin administered (−)
4. Arbutin administered (++++)
(+ or - represent relative color intensity)

In light of this, it is safe to emphasize the high quality of skin whitening substances such as aloewhite. Especially, as women consider speckles or freckles critical for their beauty, the development of aloewhite for cosmetic use is meaningful. Because the biosynthesis of melanin pigment is a way to protect against UV damage on the skin, the emphasis on skin whitening only may result in latent, dormant side-effects that could even become fatal or cause permanent damage to human. The use of aloe fortified with aloewhite has several benefits for skin whitening cosmetics because aloe has various useful functions including wound healing effect, recovering immune capability suppressed by UV irradiation (Strickland *et al.* 1994) and even moisturizing capability.

Therefore, the "RAMENT Aloe Whitening" is pre-sumably a good example of a cosmetic product expressing both skin whitening and skin care.

This research project was conducted as part of the CAP Project and was funded by the Namyang Aloe Co. From 1994 to 1997 this project was selected and funded by the Ministry of Science and Technology (MOST) of Korea as a G7 research project under the title "Screening and development of skin whitening substances". In 1994 this project was also selected as the successful G7 project by MOST of Korea, and the study results were also made public through the press by the government. The "LAMENT Aloe Whitening" was awarded the IR52 Tech-nological Achievement Award (so called, JangYoungSil Award), which was given jointly by MOST and the Korea Industrial Technology Association in 1998. This marked the first cosmetic product ever to win the award.

# References

1. Bryan BL, Fuller J, Lunsford B, Iman D (1987) Alpha-melanocyte-stimulating hormone regulation of tyrosinase in Cloudman S-91 mouse melanoma cell culture J Biol Chem 262, 4024-4033

2. Friedman PS, Gilchrest BA (1987) Ultraviolet radiation directly induces pigment production by cultured human melanocytes J Cell Physiol 133, 88-94

3. Hearing V (1987) Mammalian monophenol monooxygenase (tyrosinase): purification, properties, and reactions catalyzed. Meth. Enzymol, 142, Academic Press New York 154-165

4. Hin YH, Lee SJ, Chung MH, Park JH, Park YI, Cho TH, Lee SK (1999) Aloesin and arbutin inhibit tyrosinase activity in a synergistic manner via a different action mechanism. Arch Pharm Res 22(3):232-236

5. Maeda K, Fukuda M (1996) Arbutin: mechanism of its depigmenting action in human melanocyte culture. J Pharmacol Exp Ther 276:765-769

6. Mason H. S (1948) The chemistry of melanin. J Biol Chem 172, 83-99

7. KFDA (1994): Regulation 94-4 ; Toxicity Test Guideline

8. Pawelek JM (1976) Factors regulating growth and pigmentation of melanoma cells J Investig Dermatol 66, 201-209

9. Pawlelek JM, Korner A (1977) Activation of melanoma tyrosinase by a cyclic AMP-dependent protein kinase in a cell-free system Nature, 267, 444-447

10. Pomerantz SH (1966) The tyrosine hydroxylase activity of mammalian tyrosinase Anal Biochem 75, 86-90

11. Prota G, Crescenzi S, Misuraca G, Nicolaus RA (1970) New intermediates in pheomelanogenesis in vitro 1058-1059

12. Strickland FM, Pelley RP, Kripke ML (1994) Prevention of Ultraviolet Radiation-Induced Suppression of Contact and Delayed Hypersensitivity by Aloe barbadensis. Gel Extract J Invest Dermatol 102(1), 1-8

13. Yagi A, Kanbara T, Morinobu N (1986) Inhibition of mushroom-tyrosinase by aloe extract Planta Med 515-517

## 4.5. Cell growth-stimulating effect

### Lee, Seung Ki, Ph.D.

College of Pharmacy, Seoul National University

#### 4.5.1. Introduction

Dioscorides, a Greek physician, writing about the pharmacological effects of aloe in his book "Greek herbal" in around the first century, described his success in treating wounds, skin chapping and hair loss with aloe vera. This historical medical record from two thousands years ago already suggested that extracts of aloe vera may contain some active components with physiological effects on promoting cell growth and subsequently on regenerating human skin tissues.

These wound healing effects have been demonstrated from studies using several histological types of human and animal cells that have been used as wound healing models for both in vivo and in vitro studies. In addition, it has been reported that aloe substances contain lectin-like substances that can stimulate lymphocyte blastogenesis and induce agglutination of human and canine peripheral blood erythrocytes. Aloe also has been shown to stimulate cell-growth and to cause more rapid wound healing of skin cells in vivo. Despite these observations and the long historical use of aloe as an important folk medicine, the components in aloe responsible for the wound healing effects and their action mechanisms remain poorly understood.

In this study, we thus first aimed to screen the cell growth stimulating component in fractions from *Aloe vera* using [$^3$H]-thymidine incorporation and MTT assay. Using activity-based fractionation method and instrumental structure analysis, we were able to identify aloesin, the 4-chromone derivative that is an active ingredient responsible for exhibiting human liver cell growth stimulating effects in aloe vera.

In mammalian cells, cell growth is known to be regulated by several checkpoints in the cell cycle. For the G1/S transition, G1 kinases play important roles. Recent reports showed that G1 kinases include the cyclin complexes E-Cdk2 and D-Cdk4/Cdk6 (Winters 1993; Heggers et al. 1993). Many investigators demonstrated that the enzymatic activities of G1 kinases can be regulated by changes in the levels of cyclins, cdk, and/or by activating kinase (Hunter and Pines 1994; Sherr 1994). Furthermore, G1 kinases can be regulated by a set of proteins that stimulate cyclin dependent kinase activity through stoichiometric mechanisms (Fisher and Morgan 1994; Poon et al. 1993).

We have examined whether aloesin and methylaloesin, two chemical derivatives of aloesin, may exert a cell growth-stimulatory effect by affecting cell-cycle regulators in SK-HEP-1 cells. The present results demonstrated that aloesin and its derivative stimulate growth of SK-HEP-1 cells, which is accompanied by activation of the cyclin E-dependent kinase activity in a dose-dependent manner. This study also showed that this compound up-regulates the intracellular levels of cyclin E, cdk2, and cdc25A in the cells. Therefore, we suggest that aloesin and methylaloesin stimulate cell growth by inducing levels of cyclin E/cdk2 complex and Cdc25A, which in turn elevate the intracellular activity of cyclin E/Cdk2 kinase.

### 4.5.2. Results

### 4.5.2.1. Aloesin and methylaloesin stimulate proliferation of SK-HEP-1 cells

We have suggested that aloesin, a 4-chromone derivative, is one of the major components of *Aloe vera* and that it may have a mammalian cell growth-stimulatory effect. To determine whether aloesin (Fig 1) stimulates growth of SK-HEP-1 cells, we assessed the DNA synthetic activities in SK-HEP-1 cells in response to this agent using [$^3$H] thymidine incorporation assay method. The agent stimulated the DNA synthesis in a dose-dependent manner (Fig 2). concentration range of 1-50 $\mu$M, the DNA synthetic activity was stimulated by 2- to 4-fold by aloesin, as compared with the control values.

### 4.5.2.2. Protein synthesis is required for the stimulatory effect of aloe-in and methylaloesin on DNA synthesis

In an effort to explore the mechanism of action, we examined if the stimulatory effect of aloesin and its derivative on DNA synthesis might require protein synthesis in the cells using cycloheximide (Fig 3). The results showed that cycloheximide suppressed the DNA synthetic activities in the cells stimulated by aloesin almost back to the control values. These results indicated that the cell-growth promoting effect of these compounds is associated with newly synthesized proteins.

Since the cell-growth promoting effect of these compounds requires newly synthesized proteins, we further examined whether protein levels of cyclin E-cdk2 complex are altered in the cells after treatment with aloesin. The intracellular levels of cyclin E and cdk2 proteins were measured by immunoblotting using specific antibodies against cyclin E and cdk2 (Fig 4).

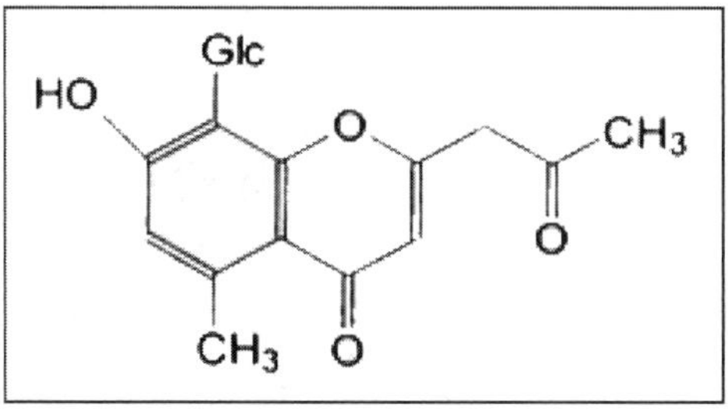

**Fig. 1.** Structure of aloesin and methylaloesin.

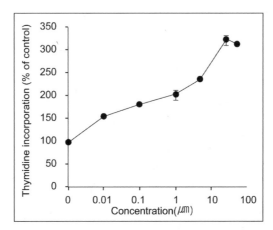

**Fig. 2.** Dose-dependent effects of aloesin and methylaloesin on DNA synthesis in SK-HEP-1 cell cultures: SK-HEP-1 cells were seeded in DMEM in the presence of 5% CS for 24 h and incubated in serum-free medium containing 1.5 mM hydroxyurea. Various concentrations of aloesin or methylaloesin were added and incubated for 24 h at 37 *C. [3H]-thymidine was added to each well for 8h. Data are expressed as percentage changes relative to the control values.   Data are presented as the mean ± SD of experiments performed in triplicate, representative of three separate experiments.

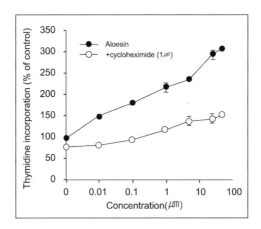

**Fig. 3.** Effects of aloesin and cycloheximide on DNA synthesis in SK-HEP-1 cell culture: SK-HEP-1 cells were seeded in DMEM in the presence of 5% CS for 24 h and then incubated in serum-free medium containing 1.5 mM hydroxyurea. Various concentrations of aloesin and 1 M cycloheximide were added and incubated for 24 h. [3H] thymidine was added to each well for 8 h and [3H] thymidine incorporation was assayed as a measure of DNA synthesis. Data are expressed as percentage changes as described in the legend of Fig. 2.

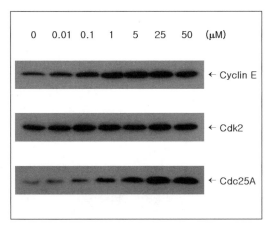

**Fig. 4.** Effects of aloesin (A) and methylaloesin (B) on the expression of cyclin E, cdk2 and cdc25A proteins in SK-HEP-1 cells: SK-HEP-1 cells were cultured in medium with increasing concentration of aloesin or methylaloesin for 24h. Cell lysates were prepared and subjected to 12% SDS-PAGE, western blotting, and immunodetection. Cyclin E, cdk2, and cdc25A proteins were detected by polyclonal anti-cyclin E, cdk2, and cdc25A antibodies, respectively, using ECL method. (See Plate 16.)

The results showed that aloesin consequently increased the levels of cyclin E and cdk2 in a dose-dependent fashion in SK-HEP-1 cells. These data suggested that these agents may up-regulate the cyclin E-dependent kinase activity by increasing the intracellular abundance of cyclin E and cdk2 proteins. Additionally, we examined whether these agents increased the intracellular level of cdc25A, which is known to play an important role in the positive regulation of cdk2 kinase activity by dephosphorylation of phosphotyrosine-15 on cdk2. The results from immunoblotting showed that the intracellular level of cdc25A increased in the cells after treatment with aloesin (Fig 4A). However, we failed to detect any changes in the level of cdc25A in the cells in response to methylaloesin (Fig 4B). These results indicated that cell-growth stimulatory effects influenced by treatment with aloesin or methylaloesin are probably due to their action on the intracellular levels of cyclin E/cdk2 complex and/or cdc25A, which may enhance the enzyme activity of cyclin E-dependent kinase in the cells.

### 4.5.2.3. Aloesin and methylaloesin up-regulate the enzyme activity of cyclin E-dependent kinase in SK-HEP-1 cell cultures

To determine the enhanced activity of Cdk2 in association with the elevated levels of cyclin E/cdk2 in the cells after treatment with these agents, we performed histone H1 kinase assay with the immuno-complexes that have been precipitated with a specific anti-cyclin E from the cell lysates.

As shown in Fig. 5, cyclin E-dependent kinase activities were significantly increased in cells treated with increasing concentrations of aloesin. Our data also showed that the stimulatory effect of aloesin was significantly higher than that of methylaloesin on cyclin E-dependent kinase activity (data not shown here), which agreed well with the effects of these agents on DNA synthesis in the cells. Taken all these results together, we propose that the cell-growth stimulatory effect of aloesin is at least partly mediated through the up-regulation of cyclin E-dependent kinase activity due to the increased level of cyclin E/cdk2 complex in the cells.

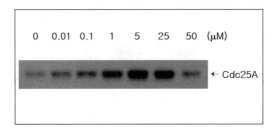

**Fig. 5.** Effects of aloesin and methylaloesin on cdk2 kinase activity in SK-HEP-1 cells: SK-HEP-1 cells were cultured in medium with increasing concentration of aloesin or methylaloesin for 24 h. Cell lysates were prepared and immune-complex kinase assays were performed using histone H1 as substrate. (See Plate 16.)

### 4.5.3. Discussion

Here we showed that aloesin, a chromone derivative in aloe vera, is the ingredient that is responsible for the cell growth stimulating activity in mammalian cell cultures, especially in SK-HEP-1 cells. Moreover, we demon-strated that the action mechanism by which this agent can stimulate the cell growth is very likely due to upregulated cyclin A/Cdk2 kinase activity that is possibly induced by elevated protein synthesis of the component proteins of this kinase.

In this study, we found that aloesin strongly stimulates the growth of SK-HEP-1 cells based on the results from [$^3$H]-thymidine incorporation (Fig 2). The growth-stimulating effects appear to require newly synthesized proteins because cycloheximide significantly blocks the stimulatory effect of these agents almost back to the control values. We then examined whether the effect of these agents is associated with increased protein levels of the cell cycle regulators, cyclin E-dependent kinase complex and cdc25A. The immunoblotting results showed that both the agents increased the intracellular levels of cyclin E and cdk2 proteins in SK-HEP-1 cells in a dose-dependent manner.

It has been reported that cyclin E, the regulatory subunit of cdk2, plays an essential and rate-limiting functional role in allowing cells to enter the S-phase of the cell cycle (Peter M and Herskowitz I 1994; Sherr CJ *at al.* 1995). Cyclin E binds to and activates cdk2 protein kinase in exponentially proliferating cells; therefore the activity of cyclin E-cdk2 complex is directly related to the abundance of the cyclin E protein (Makino K *et al.* 1973). Our hypothesis is that the cell-growth stimulatory effect of these agents is mediated through the stimulation of cyclin E-dependent kinase activity by increasing the levels of cyclin E and cdk2 in the cells. To test this hypothesis, we examined the cyclin E-dependent kinase activities that have been immunoprecipitated from the cell lysates at the time when the intracellular levels of cyclin E and cdk2 proteins were elevated in response to the agents. These agents up regulated kinase activity in the cells in a dose-dependent fashion. Therefore, we propose that the cell-growth stimulatory effects of these agents is at least partly due to their inductive effects on the protein synthesis of cyclin E and Cdk2 with subsequent activation of the intracellular activity of cyclin E-dependent kinase.

### 4.5.4. Concluding remarks

Here we demonstrated that aloesin, a chromone derivative, is the ingredient responsible for the cell growth stimulating effect of aloe vera, which has been used for several thousand years as a folk medicine for wound healing and skin regeneration. Moreover, the cell-growth stimulatory effect of aloesin and its derivative is associated with the inductions of cyclin E-dependent kinase components and its positive regulator, cdc25A. An increased level of these proteins activates cyclin E/cdk2 kinase activity and promotes DNA synthesis of cells. This study provides the first experimental evidence showing that the aloesin present in *Aloe vera* exerts a mammalian cell-growth stimulatory effect by inducing cell cycle regulator proteins. Therefore, we propose that aloe vera is an important herb medicine with a useful clinical application to wound healing and skin or tissue regeneration.

## References
1. Davis RH, Leitner MG, Russo JM (1998)J. Am. Podiatric Med Assoc 78, 60-68
2. D.C. gowda B, Neelisiddaiah V, Anjaneyalu (1979) Structural Studies of Polysaccharides from Aloe Vera. Carbohydrate Research 72: 201
3. Dulic V, Lees E, Reed SI (1992) Science 257, 1958-1961
4. Fisher RP, Morgan DO (1994) Cell 78, 713-724
5. Heggers JP, Pelley RP, Robson MC (1993) Phytother Res 7 S48-S52
6. Heggers JP, Kucukcelebi A, Stabenau CJ Ko F Broemeling LD, Robson MC, Winters WD (1993) Phytother Res 9, 455-457
7. Hoffmann I, Draetta G, Karsenti E. (1994) EMBO J 13, 4302-4310
8. Hunter T, Pines J (1994) Cell 79, 573-582
9. Jinno S, Suto K, Nagata A, Igarashi M, Kanaoka Y, Nojima H, Okazama H (1994) EMBO J 13 1549-1556
10. Makino K, Yagi A, Nishioka I (1973) Chem Pharm Bull 21, 149-156
11. Ohtsubo M, Roberts JM (1993) Science 259 1908-1912
12. Pagano M, Pepperkak R, Lukas J, Baldin V, Ansorge W, Bartek J, Draetta GJ (1993) Cell Biol 121 101-111
13. Peter M, Herskowitz I (1994) Cell 79 181-184
14. Poon RYC, Yamashita, K, Adamczewski J, Hunt T, Shuttleworth J (1993) EMBO J 12 3123-3132
15. RH Davis (1992) Inhibitory and Stimulatory Systems in Aloe Vera. Aloe TodayWinter
16. Sherr CJ (1994) Cell 79, 551-555
17. Sherr CJ, Roberts JM. Genes (1995) Dev 9, 1149-1163
18. Suzuki I, Saito H, Inous S, Migita S, Takahashi T (1979) J Biochem 89, 163-171
19. Winters WD, Benavides R, Clouse WJ (1981) Econ Bont 35, 89-95
20. Winters WD (1990) Acad Indust Joint Cong 3, 46-47
21. Winters WD (1991) Advances New Drug Develop 40, 391-397
22. Winters WD (1993) Phytother Res 7, S23-S25

## 4.6. Protective effect on nephrotoxicity

## Choung, Se Young, Ph.D.

College of Pharmacy, Kyung Hee University

### 4.6.1. Introduction

The kidney is a complex organ whose main function is to regulate the composition of body fluids. Consequently the kidney is well supplied with blood vessels and related carrier systems that transport many substances found in the body fluids. Problem arise when the kidney, in its homeostatic role vis-à-vis body fluids, is exposed to toxins.

The kidney's sensitivity to toxins is due to the following various factors:

1. The kidney receives a major proportion of the cardiac output. Consequently it is exposed to a large amount of toxins.

2. The toxic substances are concentrated in the renal tubules where electrolytes, nutrients and water are re-absorbed. As a result, the kidney is exposed to toxins at a higher concentration than other organs.

3. The proximal convoluted tubules of the kidney actively absorb glucose and amino acid while at the same time secreting metabolites like organic acid (i.e., uric acid and mercapturate) and organic bases (i.e., dopamine and creatinine). If toxic substances are re-absorbed or re-secreted during their transport through the kidney, they are concentrated in the cells of the proximal convoluted tubules and site-specific injury happens.

4. The proximal convoluted tubule and the distal convoluted tubule contain enzymes that can metabolize exogenous materials (i.e.,cytochrome P-450, cysteine conjugate β-lyase, and prostaglandin H synthase). If the metabolite is also toxic, the resultant deleterious effect of the toxic exogenous substance is greater.

Owing to this characteristic, the kidney is more frequently exposed to exogenous materials of high concentration and their metabolites than other organs are. These chemicals cause nephrotoxicity and damage the proximal convoluted tubule of the kidney which is the target organ for most nephrotoxic materials. (Hook *et al.* 1979; Rush *et al.* 1984; Hook *et al.* 1986)

Cisplatin is used in clinical practice for various cancers. It has excellent anti-cancer activity, but its use is hampered by its main side effect, namely, nephrotoxicity. Cisplatin is reported to be clinically effective against testicular cancer, ovarian cancer, bladder cancer, cervical cancer, osteogenic sarcoma, and child tumor. (Samuels *et al.* 1975; Yagoda *et al.* 1977; Wittes *et al.* 1975; Ellerby *et al.* 1974; Kamalakar *et al.* 1976) After intravenous administration, cisplatin is rapidly distributed to all the organs except the brain because of the blood brain barrier. It is therefore found in high concentration in the liver, the kidney and the intestines, but in low concentration in the brain. In the blood, 65-97% exists in the protein-bound form. (Nicolas *et al.* 1978) Its first half-life in the blood is 25~49 minutes and its second half-life, 58~73 hours. (Rosenerg *et al.* 1975)

The major toxic effects of cisplatin are as follows:

**(1) Gastrointestinal toxicity**

At a low concentration of 30mg/m$^2$ of body surface area, significant nausea and vomiting are found and anorexia can last for a week. (Lippman *et al.* 1973)

**(2) Otogenic/aural toxicity**

Ringing in the ears or tinnitus is found. Loss of hearing or deafness can also occur (Piel *et al.* 1974), and is frequently observed when patients are exposed to high doses. (Kovach *et al.* 1973)

**(3) Bone-marrow suppression:**

Bone marrow toxicity due to cisplatin is significant but transient. (Dufour *et al.* 1990) When 50~60mg/m$^2$ is administered, leukopenia, thrombocytopenia,and anemia are

observed. Leukopenia precedes the bone-marrow toxicity. (Kovach *et al.* 1973; Ronald *et al.* 1973)

In addition, cisplatin causes intravascular coagulation with capillary thrombus formation leading to thrombotic microangiopathy, microangiopathic hemolytic anemia and thrombocytopenia. (Alan *et al.* 1984) Hyperferremia, due to a complex mechanism, is also seen. Cisplatin damages reticuloendothelial cells containing iron, causing iron to be discharged into the blood. At the same time it affects the function of the erythroid precursor interfering with the entry of iron into the bone-marrow, thus causing the iron to remain in the blood in high concentration. (Sartori *et al.* 1991)

**(4) Nephrotoxicity**

The structural and biochemical characteristics of cisplatin nephrotoxicity are as follows:

The kidney has a high concentration of enzymes that are related to the synthesis and breakdown of glutathiones such as γ-glutamyl cysteine synthetase, γ-glutamyl transferase, glutathione s-transferase, and glutathione oxidase. They are maximally found in the proximal convoluted tubule, which also has a lot of renal lysosome, compared to the small amount in the distal convoluted tubule, and peroxisome. D-amino acid oxidase, L-hydroxy acid oxidase, fatty acid Co-A oxidase and choline oxidase (that makes $H_2O_2$[25]), and (OH) vitamin $D_3$-1-hydroxylase) are also found in the proximal convoluted tubule. (Walter et al. 1984)

In the body, platinum remains easily on the renal corticomedullary junction. Especially, by positive transportation, organic anions are mostly secreted so they are accumulated in the pars recta where the transported anions can easily accumulate through the same mechanism. (Robin *et al.* 1983) The accumulation of platinum in the kidney depends on energy and temperature, and only the cisplatin absorbed by the transporter of the kidney shows toxicity. (Robert *et al.* 1984) The nephrotoxicity of cisplatin is similar to that of mercury, which lies close to it in the periodic table. (Nicolas *et al.* 1978) It is said to be increased by gentamicin or other aminoglycosides. (Jon and Julian 1981) The frequency of cisplatin nephrotoxicity is 25~33% with single administration, but 50~75% with repetitive administration. The nephrotoxicity of cisplatin is not caused by platinum, but by its metabolite, aqueous form or hydroxylated complex. There are two pieces of evidence indicating that cisplatin nephrotoxicity is not caused only by platinum atoms. (Robin *et al.* 1983)

a) Uniqueness of solid structure: The kidney has the same concentration of cisplatin and transplatin, but transplatin has no nephrotoxicity, indicating that the solid structure works for it.

b) Correlation of structural activities: Once the ligand of platinum goes through metathesis, it makes great changes in the extent and frequency of nephrotoxicity.

The characteristic symptoms of cisplatin nephrotoxicity leading to acute tubular necrosis are electrolyte imbalance, raised blood urea and serum creatinine, hyperuricemia, hypomagnesemia, proteinuria, enzymuria, fall increatinine clearance, increase of urinary N-acetyl-$\beta$-glucosaminidase, leucine aminopeptidase, alanine aminopeptidase, $\beta$-glucuronidase and $\beta_2$-microglobulin, weakened re-absorptive activity of the renal tubule, necrosis of the distal convoluted tubule and the collecting tubule, prominent epithelial atypia in the collecting duct (Juan *et al.* 1977), expansion of the convoluted tubule, formation of cast, and renal failure. The latter may be fatal. (Robin *et al.* 1983)

Nephrotoxicity that is caused by cisplatin can be classified chronologically as follows:

(1) At the commencement of administration: Degenerative change in the proximal convoluted tubule, cytoplasmic vacuolization, tubular dilation, pyknotic nuclei, and hydropic degeneration.

(2) After 3~5 days: Physical change, wide tubular necrosis of the corticomedullary, pars recta necrosis of the S3 segment and the proximal convoluted tubule(Dennis *et al.* 1980), wide damage on the brush border, swelling of the cells, thickening of the nuclear chromatin, swelling of the cristae, and increase in pinocytosis.

(3) With chronic administration: Irreversible damage caused by formation of cystoma, fibrogenesis of epilepsy, and blunting of tubular basement membrane. (Dennis *et al.* 1980)

(4) Other toxic effects: Eczema, dermatitis(Robin *et al.* 1983), anaphylactic shock, facial swelling, stridor, tachycardia, hypotonia, renal failure(Rosenerg *et al.* 1975), Fanconi-type tubular dysfunction (Robin *et al.* 1983), and hemolytic uremic syndrome (HUS). HUS results in significant reduction of blood platelet, renal failure, and pulmonary edema.

A drug able to reduce the nephrotoxicity of cisplatin has not yet been developed. At present only anti-inflammatory agents and steroids are used to reduce the clinical symptoms. This research lab has recently reported that the administration of cisplatin generated lipid peroxides of the kidney and that cisplatin nephrotoxicity was associated with the formation of oxygen radicals in the immune cells of neutrophils and alveolar macrophages. The objective of this study, therefore, is to clarify the effect and mechanism of the action of prokidin, one of the ingredients of Aloe vera, in reducing cisplatin nephrotoxicity.

### 4.6.2. Main issue

In order to determine the best substance to protect the kidney from cisplatin nephrotoxicity, this study compared the effect of various natural materials and found that prokidin, extracted from Aloe vera, was the most effective. Before administering cisplatin, prokidin was administrated to Sprague-Dawley (SD) rats three times (at 72, 48 and 24 hours in advance), and the blood was collected. BUN and creatinine were then measured because cisplatin increased BUN, creatinine, proteinuria and hyperuricemia. Due to the administration of prokidin (50, 60, and 70mg/Kg), creatinine was found to return to the normal value at prokidin dosages of 60 and 70mg/Kg (Figs. 1, 2).

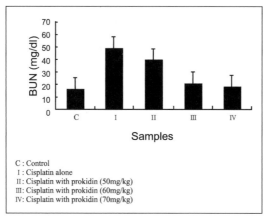

**Fig. 1.** Effects of prokidin on BUN levels in rats treated with cisplatin

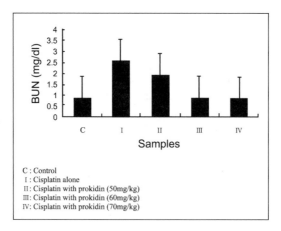

**Fig. 2.** Effect of prokidin on creatinine levels in rats treated with cisplatin

In the next experiment, the frequency of administration was changed to one time (24 hours in advance), two times (24 and 48 hours in advance) and three times (72, 48 and 24 hours in advance) with the prokidin dosage fixed at 60mg/kg.

It was found that BUN returned to the normal value with the double and triple administration, while creatinine was reduced to the normal value only with the triple administration. This result indicated that at least a triple administration is necessary to suppress cisplatin-induced nephrotoxicity (Figs. 3, 4).

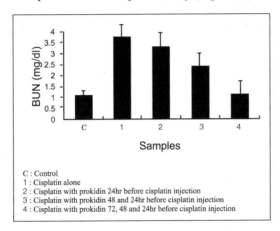

**Fig. 3.** Effects of prokidin(60mg/kg) multiple treatment on BUN levels

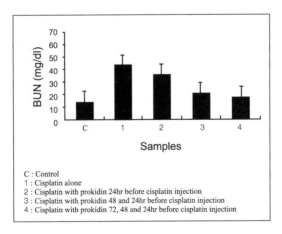

**Fig. 4.** Effects of prokidin(60mg/kg) multiple treatment on creatinine levels

With a triple administration of 60mg/kg, the time of administration was changed from 3 days before administration of cisplatin to 3 days after, and the effect on nephrotoxicity reduction was compared. The results are shown in (Fig. 5).

Suppression of cisplatin nephrotoxicity resulted in rapid weight loss and its effect was excellent with the triple administration (72, 48 and 24 hours in ad-vance). In order to rule out the possibility that prokidin's effect of reducing nephrotoxicity could also lower cisplatin's anti-cancer activity, this study examined the anti-cancer effect of administrating cisplatin only compared with that of administering both cisplatin and prokidin into ICR mice in which cancer cells had been transplanted.

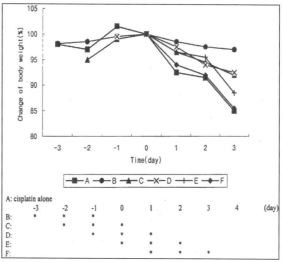

**Fig. 5.** Effects of prokidin multiple treatment on body weight change (cisplatin : prokidin = 1:5, molar ratio)

When only cisplatin was administered at a dosage possibly fatal due to ne-phrotoxicity, all the mice died by the 6th day after administration. The result was however different when both prokidin and cisplatin were administered. Even at 20 days after administration, the fatalities were remarkably few and the survival rate was also improved, confirming the general anti-cancer activity of cisplatin (Fig. 6).

These results confirmed that prokidin had no effect on cisplatin's anti-cancer action even as it significantly suppressed nephrotoxicity.

It has been established that administration of cisplatin into patients with cancer or experimental animals (SD rats) increased the formation of lipid peroxides in the kidney. Therefore, in order to examine the mechanism of prokidin's action, this study attempted to determine whether the formation of oxygen-free radicals (which could cause cisplatin nephrotoxicity) was caused by alveolar macrophages and immune cell neutrophils, and to examine the extent of this activity.

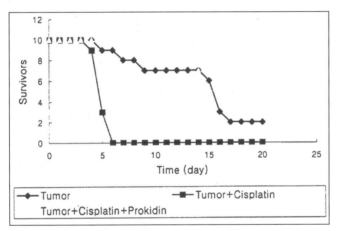

**Fig. 6.** Effects of prokidin on antitumor activity of cisplatin

In order to measure the immune cell's effect of increasing cytotoxicity, this study used neutrophils, which became activated by cisplatin, and the coculture system of LLC-PK1. The cytotoxicity of the coculture system in which cisplatin was administered was 10 times higher than that of cisplatin itself (Fig. 7).

During the experiment, the superoxide anion and hydrogen peroxide concentration increased in the coculture system, with the degree of increase depended on the culturing time. In this test, superoxide anion and hydrogen peroxide concentration were highest at 20 and 40 minutes, respectively (Figs. 8, 9).

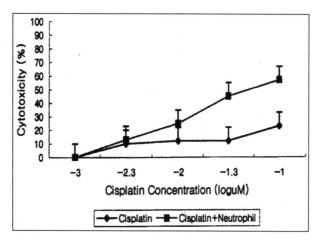

**Fig. 7.** Cytotoxicity of cisplatin increased by adding neutrophil

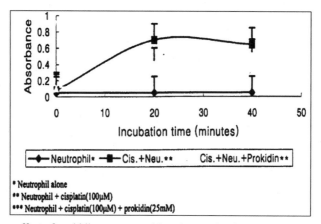

**Fig. 8.** Inhibitory effect of prokidin on superoxide anion production

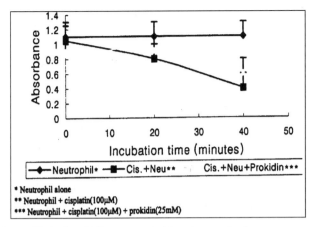

**Fig. 9.** Inhibitory effect of prokidin on superoxide anion production

As a result of administering prokidin, cisplatin cytotoxicity over neutrophils and the LLC-PK1 coculture system was reduced (Fig. 10). This reduction confirmed that the concentration of prokidin worked and, furthermore, that prokidin suppressed the neutrophils from releasing oxygen radicals, which was also dependent on the prokidin concentration.

When prokidin was administered into the coculture system in advance, the suppression of radical formation and cytotoxicity was greater than when both prokidin and cisplatin were administered together, indicating that prokidin's direct removal of oxygen radicals was weaker than its suppression over the oxygen radical formation system.

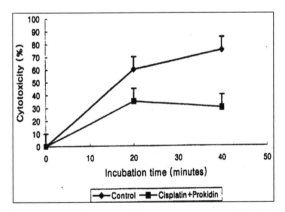

**Fig. 10.** Inhibitory effect of prokidin on cisplatin-induced cytotoxicity by inhibition of radical production

### *4.6.3. Conclusion*

The objective of this study was to assess the effect and the mechanism of action of prokidin (a chromone compound found naturally as one of the ingredients of aloe) in terms of reducing cisplatin nephrotoxicity. The results indicated that prokidin was by far the most effective suppressant of cisplatin nephrotoxicity. The optimal cisplatin dosage was found to be 3 doses of 60mg/kg each and the optimal timing to be 72, 48 and 24 hours before cisplatin administration. The absence of any deleterious influence on cisplatin's anti-cancer effect was also confirmed. The mechanism of nephrotoxicity suppression was by way of its effect on the neutrophil function of the renal blood. It stimulated the neutrophils to discharge oxygen radicals, thus preventing them from damaging organs. Based on these findings, we are of the view that prokidin has potential as a drug for reducing cisplatin-induced nephrotoxicity. However, successful efficacy trials with human subjects are required before it can be considered for clinical use.

## References

1. Alan MJ (1984) Thrombotic microangiopathy and renal failure associated with antineoplastic chemotherapy. Ann. Int. Med., 101, 41-44

2. David DC, Daniel SL (1981) Anthony, A, Acute and chronic cisplatin nephropathy in rats. Lab Kid 44, 397-402

3. Dennis CD, Joseph L, Charlotte J, Jon K, Michael WW (1980) Mechanism of cis-platinum nephrotoxicity ; II. Morphologic observations. J.Pharmacol. Exp Ther 213; 551-556

4. Dufour P, Bergerat JP, Eber M, Renayd P, Karcher V, Giron C, Leroy MJ Oberling F (1990) Cisplatin -induced acemia ; A potential interference with iron metabolism at erythroid progenitors leved. 1; 49-54

5. Ellerby RA, Davis HL Jr, Ansfield FJ (1974) Phase I clinical trial of combined therapy with 5-FU (NSC-19893) and cis-platinum(II) diamminedichloride(NSC-119875). Cancer 34; 1005-1010

6. Hook JB, McCormack M, Kluwe WM (1979) Biochemical mechanisms of nephrotoxicity. Rev. Biochem. Toxicol 1; 53-78

7. Hook JB, Hewitt WR (1986) Toxic responses of the kidney. In : Klaassen, C.D,Amdur, M.O. and Doull, J.(Eds.), Casarett and Doull's Toxicology, The Basic Science of Poisons, 3rd Edn., MAcmillan New York 310-329

8. Jon DB, Julian BH (1981) Renal and electolyte disturbances associated with cisplatin Ann Int Med 95; 628-632

9. Jon DB, Julian BH (1981) Renal and electolyte disturbances associated with cisplatin Ann Int Med 95; 628-632

10. Juan CG, Daniel MH, Esteban C, Stephen SS (1977) The renal pathology in clinical trials of cis-platin ciamminedichloride Cancer 39; 1362-1371

11. Kamalakar P, Wang JJ, Higby D (1976) Clinical Experience with cis-diamminedichloroplatinum(DDP) in chidren Proc Am Assoc Cancer Res Proc Am Soc Clin Oncol 17; 283

12. Kovach JS, Moertel CG, Schutt AJ (1973) Phase II study of cis-diamminedichloroplatinum (NSC-119875) in advanced carcinoma of the large bowel Cancer Chemother Rep 57; 357-359

13. Lippman AJ, Helson C, Helson L (1973) Clinical trials of cis – diamminedichloroplstinum (II)(NSC-119875). Cancer Chemother Rep 57; 191-200

14. Nicolas EM, John TH (1978) Platinum nephrotoxicity Am J Med 65; 307-314

15. Piel IJ, Meyer D, Perlia CP (1974) Effects of cis-diammine-dichloroplatinum(NSC- 119875) on hearing function in man. Cancer Chemother Rep 58; 871-875

16. Robert S, Peter M, Joseph BG (1984) Uptake and metabolism of cisplatin by rat kidney Kid Int 25; 753-758

17. Robin SG, Gilbert HM (1983) The nephrotoxicity of cisplatin Life Sciences 32; 685-690

18. Ronald CD, Bartlett RT, Robert CL, William AC (1973) Clinical and pharmacological studies with cis-diamminedichloroplatinum(NSC-119875). 33, 1310-1315

19. Rosenerg ML, Holoye PY, Johnson DE (1975) Bleomycin combination chemotherapy in the management of testicular neoplasia Cancer 36; 318-356

20. Rush GF, Smith JH, Newton JF, Hook JB (1984) Chemically induced nephrotoxicity ; role of metabolic activation. CRC crit Rev Toxicol 13; 99-160

21. Samuels ML, Holoye PY, Johnson DE (1975) Bleomycin combination chemotherapy in the management of testicular neoplasia Cancer 36; 318-356

22. Sartori S, Nielsen I, Masotti M, Malacarne P (1991) Early and late hyperferremia during cisplatin chemotherapy J Chemo 3; 45-50

23. Walter GG, Brian DR (1984) Enzyme distribution along the nephron Kid Int 26; 101-111

24. Wittes RE, Brescia F, Young CW (1975) Combination chemotherapy with cis-diamminedichloroplatinum(II) and bleomycin in tumors of the head and neck Oncology 32; 202-207

25. Yagoda A, Watson RC, Whitmore WF (1977) Adriamycin in advanced urinary tract cancer ; Experiene in 42 patients and review of the literature Cancer 39; 279-285

## 4.7. Immunomodulatory activity

**Lee, Chong Kil, Ph.D.**

College of Pharmacy, Chungbuk National University

### 4.7.1. Introduction

Immunomodulators are biological response modifiers that either enhance or suppress the immune responses. Immunomodulators have been used in clinical practice to treat certain cancers, viral infections, autoimmune diseases, and immunodeficiency diseases. Immunotherapy using immunomodulators is usually more effective in combination with or following chemotherapy and radiation therapy.

Aloe components exhibiting immunomodulatory activities can be divided into 2 broad categories: glycoproteins (lectins) and polysaccharides. Aloctin A may be the most well-characterized lectin isolated from aloe species, and has been shown to inhibit the growth of fibrosarcomas *in vivo* (Imanish *et al.* 1981) and to activate cytotoxicity of spleen cells and peritoneal exudates cells *in vivo* (Imanish and Suzuki 1984). Addition of aloctin A to cells *in vitro*, however, did not inhibit the growth of the cells, suggesting that the anti-tumor cell activity of aloctin A is mediated through activation of the immune responses (Imanish *et al.* 1981). In fact, aloctin A was shown to activate T cells to produce lymphokines such as IL-2 (Imanish and Suzuki 1986; Imanish *et al.* 1986; Imanish 1993). Aloctin A was also shown to enhance the production and activation of macrophages (Saito 1993).

Aloe has been shown to contain numerous types of polysaccharides such as acetylated mannan (McAnalley *et al.* 1988; Manna and McAnalley 1993; Yagi *et al.* 1977), glucomannan (Farkas 1967; Gowda *et al.* 1979), galactogalacturan (Mandal and Das 1980), and glucogalactomannan (Haq and Hannan 1981). Among the polysaccharides described from aloe species, acemannan, a mixture of various-length polymer chains of β-(1,4)-linked acetylated galactomannan, may be the most well-characterized polysaccharide isolated from the gel of *Aloe vera* [Manna and McAnalley 1993]. Acemannan has been known to have diverse immunomodulatory activities *in vivo* as well as *in vitro* (Reynolds and Dweck 1999). When administered intraperitoneally to tumor-implanted mice, acemannan was shown to cure completely or reduce the tumor burden significantly (Peng *et al.* 1991). Acemannan has also been shown to be effective in the treatment of spontaneously developed canine and feline fibrosarcomas (Harris et al. 1991; King *et al.* 1995). Immunoaugmenting activities of acemannan have also been demonstrated in numerous other systems. Acemannan has been shown to increase lymphocyte responses to alloantigens *in vitro* (Womble *et al.* 1998; Womble *et al.* 1992), exhibit adjuvant activity in vaccination against virus (Chinnah *et al.*1992) or heart-worm antigen (Usinger *et al.* 1997), and increase survival rate in virus-infected animals (Sheets *et al.* 1991; Yates *et al.* 1992).

Immunoaugmenting activity of acemannan has also been acknowledged in numerous animal models infected with viruses (Sharma *et al.* 1994; Nordgren *et al.* 1992).

The primarily focus of this chapter is the mechanism for the immunomodulatory activity of aloe polysaccharides. Then, the molecular size-activity relationship will also be discussed in an effort to clarify the optimal molecular size of aloe polysaccharides exhibiting maximum immunomodulatory activity.

### 4.7.2. Mechanisms for the immunomodulatory activity of aloe polysaccharides

The immunomodulatory activity of aloe polysaccharides appears to be mediated via activation of the host defense mechanisms, and not by the direct cytotoxicity to tumor cells. The immune response results from cascades of molecular and cellular cooperation between lymphocytes. To initiate an immune response, antigens must be trapped by professional antigen presenting cells such as macrophages and dendritic cells dendritic cells (DCs). The trapped antigens are then processed and presented to helper T cells. Activated helper T cells specifically interact with B cells that were primed with antigens. In an analogous fashion, activated helper T cells stimulate proliferation and differentiation of cytotoxic T cells. In this cell-cell cooperation, soluble mediators (i.e., cytokines) produced from the immune cells play an important role, as does direct cell-cell contact.

### 1) Activation of macrophages

The immunoaugmenting activities of aloe polysaccharides appeared to be mediated primarily through activation of professional antigen presenting cells. The immunomodulatory activities of acemannan are summarized in Table I. Acemannan was shown to activate macrophages to produce inflammatory cytokines such as IL-1, IL-6 and TNF-$\alpha$ (Womble *et al.* 1988; Womble *et al.* 1992; Zhang and Tizard 1996). Acemannan, in the presence of IFN-$\gamma$, was also shown to markedly increase NO production by macrophages (Zhang and Tizard 1996; Karaca *et al.* 1995; Djeraba and Quere 2000), and to upregulate phagocytic and candidacidal activities (Stuart *et al.* 1997). Acemannan may exert these immunomodulatory activities via binding to mannose receptors on the macro-phages (Karaca *et al.* 1995; Tietze *et al.* 1982).

**Table 1.** Immunomodulatory activity of acemannan on macrophages.

| Activity | References |
|---|---|
| Stimulation of macrophage formation | Egger *et al.*, 1996a |
| Induction of cytokines | Marshall *et al.*, 1993 |
| Stimulation of phagocytosis and   candidacidal activity | Stuart *et al.*, 1997 |
| Induction of cytokines | Tizard *et al.*, 1991 |
| Stimulation of cytokine production, nitric oxide production   and surface molecules by macrophages | Zhang and Tizard, 1996 |
| Stimulation of chicken macrophages | Djeraba and Quere, 2000 |
| Stimulation of nitric oxide production in chicken macrophages | Karaca *et al.*, 1995 |
| Induction of apoptosis in a macrophage cell line | Ramamoorthy and Tizard, 1998 |
| Stimulation of nitric oxide production in a macrophage cell line | Ramamoorthy *et al.*, 1996 |

## 2) Activation of dendritic cells

The effects of acemannan on immature DCs were recently investigated (Lee *et al.* 2001). Numerous reports in recent years have documented that DCs are the most important accessory cells for the activation of naïve T cells and the generation of primary T cell responses (Banchereau *et al.* 2000). Resident DCs, however, are least efficient in performing these functions, and must be activated and differentiated further by microenvironmental signals. Maturated and differentiated DCs are distinguished from immature DCs in that they express high levels of class II MHC molecules and co-stimulatory molecules such as B7-1, B7-2, CD40 and CD54, and exhibit strong allo-stimulatory activities (Banchereau *et al.* 2000).

To examine the effects of acemannan on immature DCs, immature DCs were generated from mouse bone marrow cells by culturing in a medium supplemented with GM-CSF and IL-4, and then stimulated with acemannan, sulfated acemannan, and LPS. The resultant DCs were examined for phenotypic andfunctional properties. Phenotypic analysis for the expression of class II MHC molecules and major co-stimulatory molecules such as B7-1, B7-2, CD40 and CD54 confirmed that acemannan could induce maturation of immature DCs (Fig. 1).

Functional maturation of immature DCs by acemannan was also supported by increased allogeneic mixed lymphocyte reaction (MLR) (Fig. 2) and IL-12 production (Fig. 3).

When sulfated chemically with chlorosulfonic acid, acemannan was unable to induce functional maturation of immature DCs. These results suggest that the immunomodulatory activity of acemannan is at least in part due to differentiation-inducing activity on immature DCs.

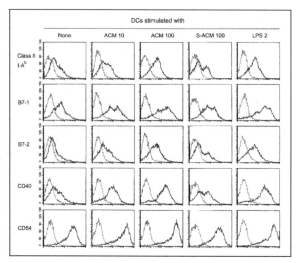

**Fig. 1**. Induction of phenotypic maturation of DCs by acemannan. Immature DCs were generated from mouse BM cells by culturing for 6 days in a medium supplemented with GM-CSF and IL-4. Immature DCs were then stimulated with acemannan (ACM, 10 and 100 μg/ml), sulfated acemannan (S-ACM, 100 μg/ml), and LPS (2 μg/ml) for 2 days. The resultant DCs were used for immunophenotypic analysis. Immunophenotypic profiles of DCs are shown compared to isotype controls.

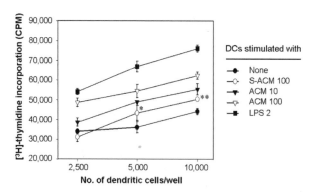

**Fig. 2.** Allostimulatory capacity of DCs cultured with acemannan. Immature DCs generated from mouse BM cells were stimulated with acemannan (ACM, 10 and 100 μg/ml), sulfated acemannan (S-ACM, 100 μg/ml) and LPS (2 μg/ml) for 2 days. The resultant DCs were treated with mitomycin C, washed, and then used as stimulators in allogeneic MLR. Proliferation of T cells in the allogeneic MLR was measured by 3H-thymidine incorporation for the final 6 hrs of the culture period of 4 days. Each data point represents the mean±SD of values obtained from three individual experiments.

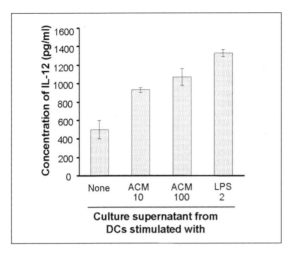

**Fig. 3.** Production of IL-12 by DCs cultured with acemannan. The culture supernatants of the DCs stimulated as described in Fig. 2 were collected and assayed for the levels of IL-12. IL-12 concentration was measured by a commercial immunoassay kit that detects the bioactive p70 heterodimer according to the manufacturer's instruction.

### 3) Effects on other immune cells

Immunomodulatory activity of acemannan may also be mediated through augmentation of hematopoiesis. Acemannan was shown to increase splenic and peripheral blood cellularity, as well as hematopoietic progenitors in the spleen and bone marrow when administered subcutaneously into myelosuppressed mice with 7 Gy radiation (Egger *et al.* 1996a). The hematopoiesis-enhancing activity of acemannan appeared to be stronger than that of granulocyte-colony stimulating factor (G-CSF), although the activity was variable depending on the doses (Egger *et al.* 1996b; Talmadge *et al.* 2004). Acemannan was also shown to increase the number of several types of leukocytes in horses (Green 1996).

Acemannan was shown to increase natural killer (NK) cell activity (Marshall and Druck 1993). The NK cell stimulatory activity of acemannan, however, may be an indirect one, because as shown in a recent study, acemannan activates DCs to produce IL-12 which activates NK cells strongly (Lee *et al.* 2001).

The immunomodulatory activity of acemannan may also be mediated through suppression of production of immunosuppressive cytokines. Acemannan was shown to prevent ultraviolet B (UVB)-suppressed immune responses in the skin (Strickland *et al.* 1994; Qui *et al.* 2000). The UVB-protective activity of aloe polysaccharides appeared to be mediated not only through prevention of apoptotic death of UVB-irradiated Langerhans cells and in the skin (Strickland *et al.* 1994), but also through inhibition of IL-10 from UVB-irradiated keratinocytes (Byeon *et al.* 1998).

### 4.7.3. Structure-activity relationship of aloe polysaccharides

Total polysaccharide content of fresh native aloe gel is approximately 10% by dry weight, and the average molecular weight of native polysaccharides in aloe gel is thought to be 2 million Da or above. The types and molecular sizes of the polysaccharides isolated from aloe gel appear very diverse (Reynolds and Dweck 1999), which may be due to differences in plant subspecies or different geographical origin. In addition, heterogeneity in molecular size may also result from technical differences used to isolate the polysaccharide, or degradation of polysaccharides by endogenous enzyme activity.

Although there is a general consensus on the diverse immunomodulatory activities of the polysaccharide fraction isolated from aloe gel, the optimal molecular size exhibiting maximum immunomodulatory activity is not clear. Acemannan, which has been shown to have multiple therapeutic properties including wound healing and immunomodulating activities, was initially reported as a β-(1,4)-acetylated mannan polymer with an average molecular weight of 80 KDa (Manna and McAnalley 1993). An injectable form of acemannan, CARN 750, developed from Carrington Laboratories (Irving, Texas), consists of acemannan polysaccharide polymers with an average MW of 1 million Da. More recently, another form of high molecular weight polysaccharide, aloeride, at between 4 and 7 million Da was isolated from aloe gel, and shown to have potent immunostimulatory activity (Pugh *et al.* 2001). The same authors claimed that the immunomodulatory activity of acemannan preparation was attributed to trace amounts of aloeride (Pugh *et al.* 2001). In contrast, a much smaller form of highly acetylated polysaccharide, termed as modified aloe polysaccharide (MAP), was isolated from cellulase-treated aloe gel, and was reported to have much stronger immunomodulatory activities than native, 2-million Da polysaccharides (Qui *et al.* 2000). MAP consists of polysaccharides with an average MW of 80 KDa. Other active polysaccharides isolated from aloe gel include a 70 KDa polysaccharide (Madis *et al.* 1989), and polysaccharides between 420 KDa and 520 KDa (Farkas and Mayer 1968).

The molecular size-immunomodulatory activity relationship of aloe polysaccharides was investigated recently with MAP. To determine the optimal molecular size of MAP exhibiting maximum immunomodulatory activity, polysaccharides were isolated from the gel of aloe *vera* that was partially digested with cellulase, and then fractionated according to their molecular size into 3 fractions, G2E1DS3 (MW ≥ 400 KDa), G2E1DS2 (5 KDa ≤ MW ≤ 400 KDa) and G2E1DS1 (MW ≤ 5 KDa), by Sephacryl column chromatography and ultrafiltration (Im et al., 2005). Examination of the immunomodulatory activities of the polysaccharide preparations showed that polysaccharides between 400 KDa and 5 KDa (G2E1D2) exhibit the most potent macrophage-activating activity as determined by increased cytokine production and nitric oxide release in Raw 264.7 cells (Fig. 4). In accordance with the *in vitro* results, examination of anti-tumor activity in an ICR strain of mouse implanted with sarcoma 180 cells showed that polysaccharides between 400 KDa and 5 KDa also exhibited the most potent antitumor activity *in* vivo (Table II).

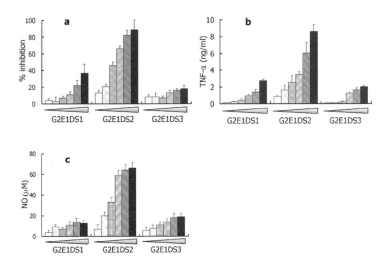

**Fig 4.** Molecular size-macrophage activating activity relationship of MAP. Protein-free MAP was separated into 3 fractions, G2E1DS1, G2E1DS2 and G2E1DS3, based on their molecular size, and then added to cultures of RAW 264.7 for 2 days. Proliferation of RAW 264.7 cells was measured by 3H-thymidine incorporation for the final 6 h (a). The culture supernatants collected after 2 days of culture were assayed for TNF-$\alpha$ (b) and nitric oxide (c). The final concentration of G2E1DS1, G2E1DS2 and G2E1DS3 in a, b and c was 0.8, 4, 20, 100, 500, and 1,000 µg/ml.

**Table 2.** Antitumor activity of MAP on sarcoma 180 cells.

| | Dose (*mg*/mouse/day) | Sarcoma 180 cells | | PEC | |
|---|---|---|---|---|---|
| | | Number (1X10⁵) | % inhibition[b] | Number (1X10⁵) | % increase[c] |
| Control | - | 3.90 | - | 1.10 | - |
| G2E1[a] | 1 | 3.59 | 7.94 | 1.41 | 128 |
| G2E1D | 1 | 0.51 | 86.92 | 4.49 | 408 |
| G2E1DS 1 | 1 | 3.32 | 14.87 | 1.68 | 152 |
| G2E1DS 2 | 1 | 0.33 | 91.53 | 4.67 | 424 |
| G2E1DS 3 | 1 | 3.28 | 15.89 | 1.72 | 156 |

[a], Aloe polysaccharides were injected into an ICR strain of mouse (i.p., 1 mg/mouse) once a day for 6 consecutive days. On day 4 from the initiation of MAP injection, sarcoma 180 cells (4X10⁵ cells/mouse) were injected into the peritoneum of MAP-treated mouse. Peritoneal exudate cells (PECs) were collected 4 days after the injection of sarcoma 180 cells, and stained with FITC-conjugated anti-CD45 monoclonal antibody. In the flow cytometric analysis, CD45-negative histogram represents sarcoma 180 cells, and CD45-positive histogram represents host cells of hematopoietic origin.

[b], Percent inhibition was calculated as follows: {(CN - TN) / CN} × 100, where CN and TN stand for the number of sarcoma 180 cells of the control group and the treated group, respectively.

[c], Percent increase was calculated as follows: {(TN - CN) / CN} × 100, where CN and TN stand for the number of PECs of the control group and the treated group, respectively.

The fact that MAP smaller than 400 KDa exhibits potent immunomodulatory activity has important implications. For *in vivo* application, smaller MW molecules are usually preferred to higher MW molecules due to bioavailability issues. MAP between 5 KDa and 400 KDa is quite a small molecule compared with polysaccharides of a few million dalton isolated from native aloe gel. In addition, the fact that MAP smaller than 400 KDa exhibited much more potent immunomodulatory activity than MAP larger than 400 KDa suggests that even partial digestion of native aloe polysaccharide increases the immunomodulatory activity. The reason for the augmentation of immunomodulatory activity by partial digestion has not been elucidated yet. One obvious finding related to this point was that partial digestion of aloe polysaccharide was not simply a reduction of a molecular size in homogenously organized polymers. Comparison of the structure between MAP and native aloe polysaccharide showed that the average molar ratio of mannose, galactose and glucose was significantly different from each other, although the major type of linkage appeared to be identical (Qui *et al.* 2000).

### 4.7.4. Conclusion

There have been numerous reports demonstrating that polysaccharides contained in the gel of aloe species exhibit immunomodulatory activity *in vivo* as well as *in vitro*. Studies on the efficacy and toxicity of the aloe polysaccharides have even lead to the development of an injectable form of polysaccharide, acemannan (CARN 750). The mechanisms for the immunomodulatory activity has become clearer in recent years. The fact that acemannan induces phenotypic and functional maturation of immature DCs has an important implication in understanding the antiviral and antitumoral activities of acemannan, because it has been well-documented that DCs are the most important accessory cells for the activation of naïve T cells and generation of primary T cell responses. Although there is a general consensus on the diverse immunomodulatory activities of the polysaccharide fraction isolated from aloe gel, the optimal molecular size exhibiting maximum immuno-modulatory activity has been a matter of debate. Optimal molecular size exhibiting maximal immunomodulating activity may vary depending on the examination parameters for the immunomodulatory activities. Examination of the molecular size-activity relationship showed that polysaccharides between 5 – 400 KDa exhibit the most potent immunomodulatory activity. The fact that MAP smaller than 400 KDa exhibits potent immunomodulatory activity has important implications, because, for *in vivo* applications, smaller MW molecules are usually preferred to higher MW molecules due to bioavailability issues. Further studies using different *in vivo* systems are still required to establish the molecular size-activity relationship more clearly, and this type of study would be essential to develop aloe polysaccharides as therapeutic drugs.

# References

1. Banchereau J, Briere F, Caux C, Davoust J, Lebecque S, Liu YJ, Pulendran B, and Palucka K (2000) Immunobiology of dendritic cells. Annu Rev Immunol 18: 767-811.

2. Byeon SW, Pelley RP, Ullrich SE, Waller TA, Bucana CD, and Strickland FM (1998) Aloe barbadensis extracts reduce the production of interleukin-10 after exposure to ultraviolet radiation J Invest Dermatol 110: 811-7.

3. Chinnah AD, Baig MA, Tizard IR, and Kemp MC (1992) Antigen dependent adjuvant activity of a polydispersed beta-(1.4)-linked acetylated mannan (acemannan) Vaccine 10: 551-557.

4. Djeraba A., and Quere P (2000). In vivo macrophage activation in chickens with Acemannan, a complex carbohydrate extracted from Aloe vera Int J Immunopharmacol 22: 365-372.

5. Egger S, Brown GS, Kelsey LS, Yates KM, Rosenberg LJ, and Talmadge JE (1996a) Studies on optimal dose and administration schedule of a hematopoietic stimulatory -(1,4)-linked mannan Int J Immunopharmacol 18: 113-126.

6. Egger S, Brown GS, Kelsey LS, Yates KM, Rosenberg LJ, and Talmadge JE (1996b) Hematopoietic augmentation by a beta-(1,4)-linked mannan Cancer Immunol Immunother 43: 195-205.

7. Farkas A, and Mayer RA (1968) Polysaccharide from the juice of the aloe plant for treatment of wounds U.S. Patent 3,362,951.

8. Farkas A (1967) Methylated polysaccharide and method of making. US Patent, 3,360,510. Fujita, K., Suzuki, I., Ochiai, J., Shinpo, K., Inoue, S., and Saito, H. 1978. Specific reaction of aloe extract with serum proteins of various animals Experientia 34: 523-524.

9. Gowda DC, Neelisiddaiah B, Anjaneyalu YV (1979) Structural studies of polysaccharides from Aloe vera Carbohyd Res 72: 201-205.

10. Green P (1996) Aloe vera extracts in equine clinical practice. Veterinary Times 26: 16-18 Haq Q.H. and Hannan A 1981 Studies on glucogalactomannan from the leaves of Aloe vera, Tourn.(ex Linn). Bangladesh J Sci Indus Res 16: 68-72.

11. Harris C, Pierce K, King G, Yates KM, Hall J, Tizard I (1991) Efficacy of acemannan in treatment of canine and feline spontaneous neoplasms Molec Biother 3: 207-213.

12. Im AA, Oh ST, Song S, Kim MR, Kim DS, Woo SS, Jo TH, Park YI, Lee CK (2005) Identification of optimal molecular size of modified Aloe polysaccharides with maximum immunomodulatory activity Int Immunopharmacol 5: 271-279.

13. Imanishi K (1993) Aloctin A, an active substance of Aloe arborescens Miller as an immunomodulator Phytother Res 7: S20-S22.

14. Imanishi K, Ishiguro T, Saito H, and Suzuki I (1981) Pharmacological studies on a plant lectin, Aloctin A. I. Growth inhibition of mouse methylcholanthrene induced fibrosarcoma (Meth A) in ascites form by Aloctin A Experientia 37: 1186-1187.

15. Imanishi K, and Suzuki I (1984) Augmentation of natural cell-mediated cytotoxic reactivity of mouse lymphoid cells by Aloctin A Int J Immunopharmacol 6: 539-543.
16. Imanishi K, and Suzuki I (1986) Induction of nonspecific cell mediated cytotoxic reactivity from non-immune spleen cells treated with Aloctin A Int J Immunopharmacol 8: 781-787.
17. Imanishi K, Tsukuda K, and Suzuki I (1986) Augmentation of lymphokine-activated killer cell activity in vitro by Aloctin A Int J Immunopharmacol 8: 855-858.
18. Karaca K, Sharma JM, and Nordgren R (1995) Nitric oxide production by chicken macrophages activated by acemannan, a complex carbohydrate extracted from Aloe vera Int J Immunopharmacol 17: 183-188.
19. King GK, Yates KM, Greenlee PG, Pierce KR, Ford CR, McAnalley BH, Tizard IR (1995) The effects of acemannan immunostimulant in combination with surgery and radiation therapy on spontaneous canine and feline fibrosarcomas J Am Anim Hosp Assoc 31: 439-447.
20. Lee JK, Lee MK, Yun YP, Kim Y, Kim JS, Kim YS, Kim K, Han SS, Lee CK (2001) Acemannan purified from Aloe vera induces phenotypic and functional maturation of immature dendritic cells Int Immunopharmacol 1: 1275-1284.
21. Mandal G, Das A (1980) Structure of the glucomannan isolated from the leaves of Aloe barbadensis Miller Carbohyd Res 87: 249-256.
22. Madis VH, Omar MM, Madi SV (1989) Aloeferon isolation, manufacturing and its applications U.S. Patent 4,861,761.
23. Manna S, McAnalley BH (1993) Determination of the position of the O acetyl group in $\beta$-(1-4) mannan (acemannan) from Aloe barbadensis Miller. Carbohyd Res 241: 317-319.
24. Marshall GD, Druck JP (1993) In vitro stimulation of NK activity by acemannan J Immunol 150: 241A.
25. Marshall GD, Gibbons AS, Parnell LS (1993) Human cytokines induced by acemannan J Allergy and Clin Immunol 91: 295.
26. McAnalley BH, McDaniel HR, Carpenter RH (1988) Demonstration of in vitro antiviral action of acemannan (ACE-M) against multiple viruses including the HIV virus Proceedings of the IV International Conference on AIDS p158.
27. Nordgren RM, Stewart-Brown B, Rodeberg JH (1992) The role of acemannan as an adjuvant for Marek's disease vaccine Proceedings of the XIX Worlds Poultry Congress 165-169.
28. Peng SY, Norman J, Curtin G, Corrier D, McDaniel HR, Busbee D (1991) Decreased mortality of Norman Murine Sarcoma in mice treated with the immunomodulator Acemannan Mol Biother 3: 79-87.
29. Pugh N, Ross SA, EiSohly MA, Pasco DS (2001) Characterization of aloeride a new polysaccharide from Aloe vera with potent immunostimulatory activity J Agric Food Chem 49: 1030-1034.
30. Qui Z, Jones K, Wylie M, Jia Q, Orndorff S (2000) Modified Aloe barbadensis polysaccharide with immunomodulatory activity Planta Med 66: 152-156.

31. Ramamoorthy L, Kemp M.C, Tizard IR (1996) Acemannan, a-(1,4)-acetylated mannan, induces nitric oxide production in macrophage cell line Raw 264 7 Mol Pharmacol 50: 878-884.

32. Ramamoorthy L, Tizard IR (1988) Induction of apoptosis in a macrophage cell line RAW 264.7 by acemannan, a-(1,4)-acetylated mannan. Mol Pharmacol 53: 415-421.

33. Reynolds T, Dweck AA (1999) Aloe vera gel leaf: a review update J Ethnopharmacol 68: 3-37.

34. Saito H, Ishiguro T, Imanishi K, Suzuki I (1982) Pharmacological studies on a plant lectin aloctin A. II. Inhibitory effect of aloctin A on experimental models of inflammation in rats Jpn J Pharmacol 32: 139-142.

35. Sharma JM, Karaca K, Pertile T (1994) Virus-induced immunosuppression in chickens Poultry Sci 73: 1082-1086.

36. Sheets MA, Unger BA, Giggleman GR. Jr, Tizard IR (1991) Studies of the effect of acemannan on retrovirus infections: Clinical stabilization of feline leukemia virus-infected cats Mol Biother 3: 41-45.

37. Strickland FM, Pelley RP, Kripke ML (1994) Prevention of ultraviolet radiation- induced suppression of contact and delayed hypersensitivity by Aloe barbadensis gel extract J Invest Dermatol 102: 197-204.

38. Stuart RW, Lefkowitz DL, Lincoln JA, Howard K, Gelderman MP, Lefkowitz SS (1997) Upregulation of phagocytosis and candidacidal activity of macrophages exposed to the immunostimulant acemannan Int J Immunopharmacol 19: 75-82.

39. Stuart RW, Lefkowitz DL, Lincoln JA, Howard K, Gelderman MP, Lefkowitz SS (1997) Upregulation of phagocytosis and candidacidal activity of macrophages exposed to the immunostimulant, acemannan Int J Immunopharmacol. 19: 75-82.

40. Talmadge J, Chavez J, Jacobs L, Munger C, Chinnah T, Chow JT, Williamson D, Yates K (2004) Fractionation of Aloe vera L. inner gel, purification and molecular profiling of activity Int Immunopharmacol 4: 1757-1773.

41. Tietze C, Schlesinger PH, and Stahl P (1982) Mannose-specific endocytosis receptor of alveolar macrophages: demonstration of two functionally distinct intracellular pools of receptor and their roles on receptor cycling J Cell Biol 92: 417-424.

42. Tizard I, Carpenter RH, Kemp M (1991) Immunoregulatory effects of a cytokine release enhancer (Acemannan). International Congress of Phytotherapy 1991 Seoul Korea p68.

43. Usinger WR (1997) A comparison of antibody responses to veterinary vaccine antigens potentiated by different adjuvants Vaccine 15: 1902-1907.

44. Womble D, Helderman JH (1988) Enhancement of allo-responsiveness of human lymphocytes by acemannan (Carrisyn) Int J Immunopharmacol 10: 967-973.

45. Womble D, Helderman JH (1992) The impact of acemannan on the generation and function of cytotoxic T-lymphocytes Immunopharmacol Immunotoxicol 14: 63-77.
46. Yagi A, Makino K, Nishioka I, Kuchino Y (1977) Aloe mannan, polysaccharide from Aloe arborescens var natalensis Planta med 31: 17-20.
47. Yates KM, Rosenberg LJ, Harris CK, Bronstad DC, King GK, Biehle GA, Walker B, Ford CR, Hall JE, Tizard IR (1992)   Pilot study of the effect of acemannan in cats infected with feline immunodeficiency virus Vet Immunol Immunopathol 35: 177-189.
48. Zhang L, Tizard IR (1996) Activation of a mouse macrophage cell line by acemannan: the major carbohydrate fraction from Aloe vera gel Immunopharmacol 35: 119-128.

# 5. Quality control and standardization of Aloe products

## Kim, Kyeong Ho, Ph.D.[1] and Park, Jeong Hill, Ph.D. [2]

## 5.1. Introduction

Aloe species have been used for a long time in folk medicine for the treatment of constipation, burns and dermatitis. Several anthraquinones, anthrones, chromones, and their C-glycosyl compounds have been isolated from various Aloe species (Haynes and Holdsworth 1970; Makino *et al.* 1974; Speranza *et al.* 1986a, b; Mebe 1987;Conner *et al.* 1989; Park *et al.* 1995, 1996, 1997; Okamura *et al.* 1996). Aloe barbadensis (syn. A. vera) and A. arborescens are widely used in Korea as ingredients of health foods and cosmetics. Freeze-dried and hot air-dried aloe powder are the two types of available sources. However, the quality control methods for aloe or aloe preparations have not been extensively studied. Many attempts have been reported for the determination of chemical constituents of aloe, such as fluorophotometry (Ishii *et al.* 1984), thin-layer chromatography (TLC; Wawrznowicz *et al.* 1994), gas chromatography (GC; Nakamura *et al.* 1989; Nakamura and Okuyama 1990) and high performance liquid chromatography (HPLC; Suzuki *et al.* 1986; Rauwald and Beil 1993a, b; Rauwald and Sigler 1994; Zonta *et al.* 1995). However, the HPLC method can be applied directly to analyse the chemical constituents of aloe without any modification. Although some reverse-phase liquid chromatographic methods for the analysis of a few phenolic compounds have been reported, these mainly focused on the chemotaxonomic data of C-glycosyl anthrones, their profiles, chromatographic patterns, or identification of aloin A and B in aloe powders.

Thirteen phenolic components (Fig. 1), namely aloesin, 8-C-glucosyl-7-O-methyl-(S)-aloesol, neoaloesin A, 8-O-methyl-7- hydroxyaloin A and B, 10-hydroxyaloin A, isoaloeresin D, aloin A and B, aloeresin E, aloe-emodin, aloenin and aloenin B in A. barbadensis and A. arborescens were separated and quantified by the HPLC method.

The contents of aloesin in the aloe preparations were determined by HPLC. A new and sensitive method for the quantitation of trace amounts of aloesin in biological fluid by HPLC technique was developed, as was a new method to detect the adulteration of commercial aloe gel powders.

---

[1]College of Pharmacy, Kangwon National University
[2]College of Pharmacy,Seoul National University

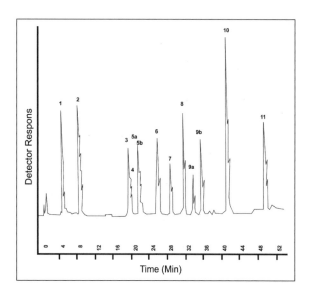

**Fig. 1.** HPLC chromatogram of a standard mixture of the thirteen phenolic compounds identified in *Aloe* species. Key to peak identity: **1.**aloesin; **2.**8-*C*-glucosyl-7-*O*-methyl-(*S*)-aloesol; **3.**neoaloesin A; **4.**aloenin; **5a.**8-*C*-methyl -7-hydroxyaloin A; **5b.**8-*C*-methyl-7 -hydroxyaloin B; **6.**aloenin B; **7.**10-hydroxyaloin A; **8.**isoaloeresin D; **9a.**aloin A; **9b.**aloin B; **10.**aloeresin E; **11.**aloe-emodin.

## 5.2. Analysis of 13 Phenolic Compounds in Aloe species by HPLC

Owing to the various polarities of the thirteen phenolic compounds, a gradient elution was employed for the simultaneous separation and determination. A methanol : water gradient gave better resolution than an acetonitrile : water gradient.

Figure 1 shows a chromatogram of the thirteen standard compounds. All of these compounds were baseline-resolved, except for aloenin and neoaloesin A, which, although found in A. arborescens and A. barbadensis independently, were not found together in the same sample.

Figures 2 and 3 show the HPLC chromatogram of extracts of A. barbadensis and A. arborescens, respectively. Chromones such as aloesin, 8-C-glucosyl-7-O-methyl-(S)-aloesol and isoaloeresin D were abundant in A. barbadensis, while pyrones such as aloenin and aloenin B were abundant in A. arborescens. Aloin A and aloin B were the most of abundant compounds in both species, whilst 10-hydroxyaloin A and aloe-emodin were also found in both species.

Figures 4 and 5 show the seasonal variation of phenolic constituents in A. barbadensis and A. arborescens in Korea. The contents of aloin A and B, 8-O-methyl-7-hydroxyaloin A and B, and 10-hydroxyaloin A in A. barbadensis (Fig. 5) showed large variations, while chromones, such as aloesin and neoaloesin A, did not. The content of 8-C-glucosyl-7-O- methyl-(S)-aloesol showed increased levels in the spring and summer, and decreased levels in the autumn and winter.

In the case of A. arborescens (Fig. 6), the contents of aloenin, aloin A and B, and 10-hydroxy aloin A, were quite consistent throughout the year, except in February. The content of aloenin B was lowest in March, and increased thereafter.

Tables 1 and 2 summarize the average content of the chemical constituents.

Aloin (or barbaloin) is known as the main laxative component of aloe preparations, and it has generally been used as a key component for the quality control of pharmaceuticals containing aloe (Ishii et al. 1984; Zonta *et al.* 1995). Recently, some species of aloe, including A. barbadensis, have also been widely used, not only as laxatives, but also as ingredients of health foods and cosmetics. In this case, it would be better to adopt aloesin, a C-glycosyl chromone, as the key component with respect to quality control of such preparations. Aloesin shows less seasonal variation than aloin, and is also more stable and resistant toward hydrolysis (by acid or base) and to high temperatures (Haynes and Holdswirth 1970; Graf and Alexz 1980). Moreover, aloin has also been reported in other plants (Fairbairn and Simic 1960), while aloesin has not been found in any other plants except Aloe species (Fianz and Gruen 1983) and can be easily analysed (Kim *et al.* 1996) In the case of A. arborescence, it would be more beneficial to use aloenin as the key component for quality control.

This method would be useful in the quantitative and quantitative analysis of the major compounds in Aloe species. It would also be useful for analyzing bulking samples, and for the quality control in the cosmetics, pharmaceutical or health food industries.

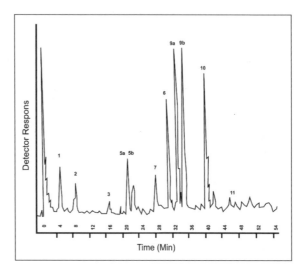

**Fig. 2.** HPLC chromatogram of *A. barbadensis*

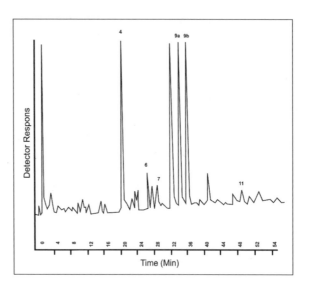

**Fig. 3.** HPLC chromatogram of *A. arborescence*

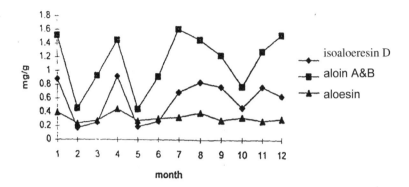

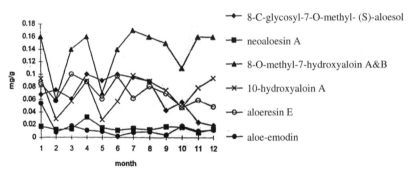

**Fig. 4.** Seasonal variation in the concentrations of phenolic components in samples of *Aloe barbadensis.*

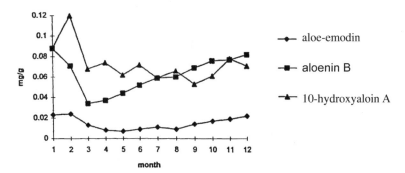

**Fig. 5.** Seasonal variation in the concentrations of phenolic components in samples of *Aloe arborescence.*

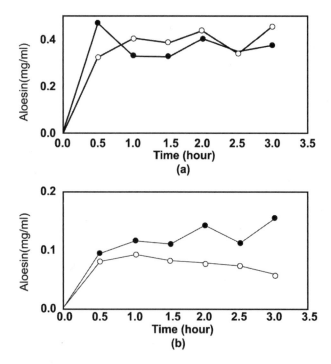

**Fig. 6.** Effect of method and time on the extraction of aloesin.
((a) sample 1 and (b) sample 2)
○ : extract with ultrasonic cleaner
● : extract with reflux

**Table 1.** Average contents of phenolic compounds in *Aloe barbadensis*

| Compound | Content(mg/g)[a] |
|---|---|
| Aloesin | 0.32±0.06 |
| 8-*C*-glucosyl-7-*O*-methyl-(*S*)-aloesol | 0.061±0.031 |
| neoaloesin A | 0.015±0.006 |
| 8-*C*-methyl-7-hydroxyaloin A& B | 0.14±0.04 |
| 10-hydroxyaloin A | 0.070±0.024 |
| isoaloeresin D | 0.57±0.27 |
| aloin A & B | 1.14±0.39 |
| aloeresin E | 0.071±0.018 |
| aloe-emodin | 0.014±0.009 |

[a]Mean (±standard deviation) : n=3.

**Table 2.** Average contents of phenolic compounds in *Aloe arborescence*

| Compound | Content(mg/g)[a] |
|---|---|
| aloenin | 1.53±0.25 |
| aloenin B | 0.062±0.016 |
| 10-hydroxyaloin A | 0.073±0.017 |
| aloin A & B | 1.21±0.25 |
| aloe-emodin | 0.015±0.0058 |

[a]Mean (±standard deviation) : n=3.

## 5.3. Determination of aloesin in aloe preparations by HPLC

Two types of commercial aloe gel powders were used as samples: freeze-dried Aloe vera gel, which was acquired by peeling of Aloe vera whole leaves (sample 1), and freeze-dried Aloe vera whole leaves without peeling (sample 2). HPLC method was developed to determine the aloesin, Chromone-*C*-glycoside, as an indicator of aloe species applied to the assay of the commercial aloe preparations.

When the reflux extraction and ultrasonic extraction were compared for the extraction of aloesin from aloe preparations, there was no significant difference in extraction yield between them, but in the case of reflux extraction, there was some interference peaks at the HPLC chromatogram (Fig. 6). Thus aloesin was extracted 3 times for 30min. with ethanol using the ultrasonic extraction method.

Ethanol extract was further purified using liquid-liquid extraction method before injection to the HPLC system. Nonpolar interferences were removed into dichloromethane and ethylacetate layers and polar interferences into water layer (Table 3). The remained n-butanol layer was evaporated to dryness and reconstituted with mobile phase and injected into the HPLC system.

The ODS column was used as a stationary phase and gradient elution from 10% methanol to 70% methanol was used as a mobile phase monitored at UV 293nm (Fig. 7). The recovery of aloesin was 98.0~123.0%.

The aloesin content in the freeze-dried aloe gel was 0.09%~0.13ng/g (Table 5). Moreover, this method could be applied to the commercial tablets and capsules.

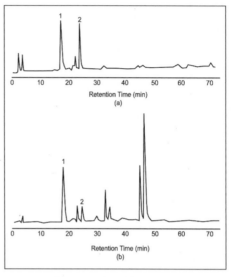

**Fig. 7.** HPLC chromatograms of (a) sample 1 and (b) sample 2. Peak 1 indicates puerarin that was used as an internal standard.

**Table 3.** Relative distribution of aloesin and ethanol ext. in dichloromethane, ethylacetate, n-butanol and aqueous fractions.

| Fraction | Sample 1 | | Sample 2 | |
|---|---|---|---|---|
| | aloesin(%) * | extracts(%)** | aloesin(%) | extracts(%) |
| Dichloromethane | 1.2 | 1.6 | 2.2 | 19.8 |
| Ethylacetate | 6.4 | 8.8 | 6.7 | 24.3 |
| n-Butanol | 79.8 | 24.0 | 78.0 | 13.1 |
| $H_2O$ | 12.6 | 65.5 | 13.0 | 42.3 |

$$* \frac{\text{weight of aloesin each solvent}}{\text{total weight of aloesin ethanol extract}} \times 100$$

$$** \frac{\text{weight of extract partitioned to each solvent}}{\text{total weight to ethanol extract}} \times 100$$

**Table 4.** Relative distribution of aloesin in dichloromethane, n-butanol and NaCl saturated aqueous fractions.

| Fraction | Sample 1 | Sample 2 |
|---|---|---|
| | aloesin(%)* | aloesin(%) |
| Dichloromethane | N.D. | N.D. |
| n-Butanol | 97.5 | 98.3 |
| NaCl aq | N.D. | N.D. |

**Table 5.** Contents of aloesin in various aloe preparations

| Sample name | Amount measured(mg/g) |
|---|---|
| Freeze-dried powder of Aloe vera gel | 0.09 |
| Freeze-dried powder of Aloe arborescence   gel | 0.13 |
| Freeze-dried  powder  of  Aloe  arborescence whole leaf | 0.59 |
| Commercial Tablet A | 0.40 |
| Commercial Tablet B | 0.31 |
| Commercial Capsule C | 0.007 |

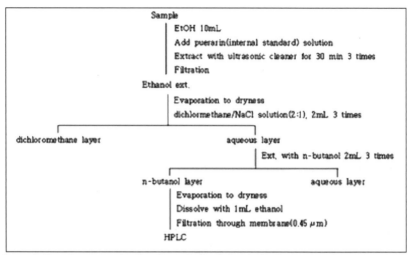

**Scheme 1.** Sample preparation for the analysis of aloesin in aloe preparations.

## 5.4. A sensitive HPLC method for the assay of aloesin in plasma

Aloesin has been reported to have various biological activities, along with a wound healing effect, anti-gastric ulcer effect, whitening effect and reducing effect of toxicity of antineoplastics (Yagi *et al.* 1987b). For the pharmacokinetic study of aloesin, a sensitive HPLC (Ed- you have already introduced this abbreviation in 5.1 introduction) method for the separation and quantitation of aloesin in plasma was developed. 9-Anthroylnitrile was used as a fluorescent labelling reagent for the analysis of aloesin. Aloesin reacted with a molar excess of 9-anthroylnitrile in the presence of quinuclidine to form a stable adduct.

9-Anthroylnitrile and quinuclidine were added to aloesin and incubated at 100°C for 5min. After incubation, methanol was added immediately to terminate the reaction. Aloesin-9-AN derivative was extracted using Sep-pak C18 cartridge and analyzed by HPLC. The reaction rate was significantly influenced by the reaction temperature and time where the maximum yield was obtained with incubation at 100°C for 5min (Fig. 9). The product was not decomposed after up to 24hrs at room temperature (Fig. 10).

For an internal standard, compound AD-1(8-*C*-β-D-glucopyranosyl-7- hydroxy -5-methyl-2-(2-hydroxypropyl)-4-H-benzopyran-4-one) was synthesized (Fig. 11).

Aloesin and internal standard (compound AD-1) were added to blank human plasma and derivatized with 9-anthroylnitrile followed by applying to the Sep-pak C18 cartridge and determined by HPLC (stationary phase, Inertsil ODS-3 (4.6×150mm, 5µm); mobile phase, acetonitrile/methanol/H$_2$O (3/1/6,v/v/v)).
The detection limit of 9-anthroyl ester of aloesin as measured by fluorescence detection was 3.2ng/mL, which was about 80 times lower than that of aloesin itself as measured at UV 293nm (Fig 12).

This method could be applied to the pharmacokinetic studies of aloesin.

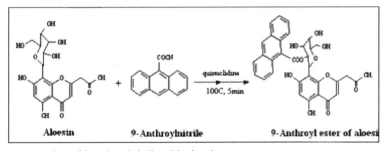

**Fig. 8.** Reaction of 9-anthroylnitrile with aloesin

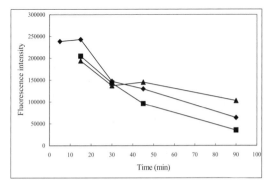

**Fig. 9.** Influence of reaction temperature and time on the derivatization of aloesin with 9-anthroylnitrile (. (100.), . (80.) and . (60.)).

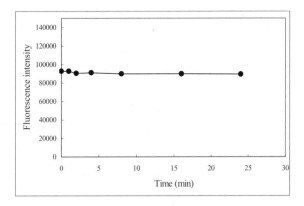

**Fig. 10.** Stability of 9-anthroyl derivatized aloesin

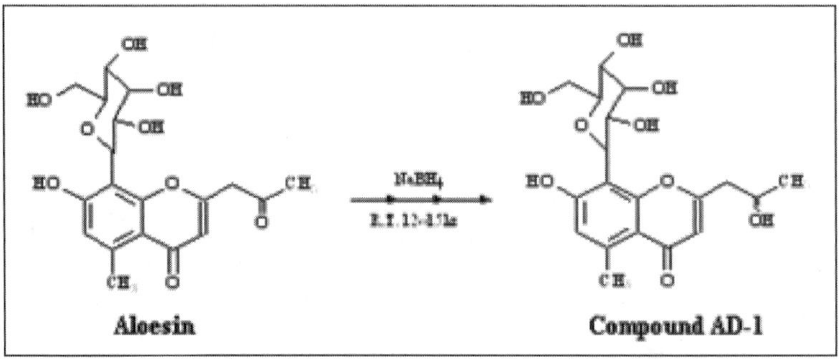

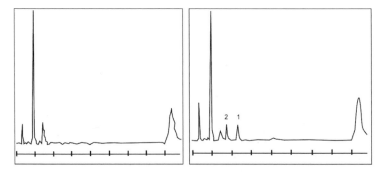

**Fig. 11.** NaBH$_4$ reduction of aloesin into compound AD-1.

**Fig. 12.** Typical chromatograms of extracts of (1) blank plasma, and (2) 150ng/ml standard aloesin spiked to plasma. Column, Inertsil ODS-3 (5um, 4.6×150mm); mobile phase, acetonitrile/methanol/d.d.w.=3/1/6; flow rate, 1.0ml/min; detection, fluorescence 360nm for excitation and 460nm for emission. Peak 1, derivative of aloesin; peak 2, derivative of internal standard.

## 5.5. Detection of adulteration of commercial Aloe Gel Powder

Recently considerable attention has been directed to the use of aloe species as healthy foods and new drugs and sales of products derived from aloe species are growing rapidly. Large numbers of aloe products are commercially applied in various fields of industry, and therefore many kinds of aloe gel powders are merchandised. Nevertheless, as in all food industries, where opportunities exist to realize large margins unscrupulous people will exploit the situation. To protect the image of Aloe as a safe and pure product, quality control of aloe products is essential. After L-malic acid was proposed as a marker of aloe products (Pelley *et al.* 1992, 1993), artificial addition of synthetic D-malic acid to the adulterated aloe products became detectable, but not of L-malic acid. We tried to determine the quality of phenolic compounds in aloe species such as aloesin, methylaloesol, aloeresin D, barbaloin and aloe-emodin etc., but were unable to because their contents varied significantly according to the processes of manufacturing.

In order to detect the adulteration of aloe product with polysaccharide, many researchers have developed various methods. The carbohydrate pattern in aloe plant was studied (Mandel *et al.* 1980a, 1980b). Polysaccharides were isolated by alcohol precipitation method and determined by the phenol-sulfuric acid colorimetric method (Hodge *et al.* 1962; Whistler *et al.* 1962). The isolated crude polysaccharides were purified by dialysis across a membrane with a cutoff of approximately 5,000 MW for removal of small molecular interfering substances. They were hydrolyzed and the resultant free sugars were determined (Gowda *et al.* 1979; Waller *et al.* 1994).

However, the analysis of polysaccharide in aloe products cannot completely reveal the presence or level of adulteration.

We have found that some commercial aloe products were adulterated with maltodextrin. The qualitative TLC method and quantitative HPLC method were developed and applied to detect such adulteration.

Standard aloe gel powder samples were prepared. Aloe vera gel which was acquired by peeling of Aloe vera whole leaves was stirred well with glass rod after cellulase had been added. The fresh, thawed, crude gel was filtered with a sieve (pore size~180 um). After adding charcoal to a filtered gel solution, the mixture was heated at 70°C for 30 min and filtered with high vacuum. The filtered solution was lyophilized and aloe gel powder was consequently acquired. Differently, whole Aloe vera gel was lyophilized and dried crude Aloe vera gel was obtained. The resultant Aloe vera gel powder and dried crude Aloe vera gel were designated as 'Sample A' and 'Sample B' respectively.

Samples A and B and commercial aloe gel powder products were mixed with ethyl alcohol until homogeneity. They were placed at 4° C for precipitation overnight. After precipitation, the contents were centrifuged at 2000 rpm for 10 min. After centrifugation, the supernatant was discarded and each precipitation was lyophilized for 24hrs. The weight of each precipitate was measured as alcohol precipitates (Table 6-c). Hexose in the alcohol precipitates was measured by phenol-sulfuric acid method (Hodge *et al.* 1962; Whistler *et al.* 1962) (Table 6-d). Each aloe gel powder (300mg) was transferred into a crucible and then heated at 500°C for 4 hrs. After cooling to room temperature in a desiccator, the weights of the residues (ash contents) were measured precisely (Table 6-b). Dialysis was carried out using a microporous membrane to separate molecules at approximately a 5,000 MW cutoff (Table 6-e). The lyophilized retained materials were placed into a hydrolysis tube and 6M-hydrochloric acid was added. After dissolution, the hydrolysis tube was evacuated of oxygen by high vacuum while being chilled in an ice bath to minimize loss of hydrochloric acid. The tube was sealed and hydrolysis accomplished by incubation for 30 min at 120°C. The solution was neutralized with 2M-sodium hydroxide solution. Each solution was transferred into a reaction vial and was lyophilized. The residue was dissolved in 1pyridin anhydrous, followed by the addition of hexamethyldisilazane (HMDS) and trimethylchlorosilane (TMCS). After mixing, the whole mixture was kept at room temperature for 5 min and then each 2-ul aliquot of solution was injected into gas chromatography and the contents of mannose, galactose and glucose were measured (Table 6-g, Fig. 13).

The contents of alcohol precipitable solid, total alcohol precipitable hexose and ash in the aloe gel powder of the standard sample A were 46.4%(w/w), 4.5%(w/w) and 23.2%(w/w), respectively. In standard sample B, the products K1, A1~A9, T1, F1 and M1 showed a similar pattern to that in the standard sample A. However, for products G1~G3, T2~T4, D1 and P1, the amounts of ash were very low while those of total alcohol precipitable hexoses were very high. So, these products were considered to be adulterated with commercial plant poly-saccharides. Relative sugar compositions in alcohol precipitable poly-saccharides were investigated using gas chromatography. Contents of mannose, galactose and glucose in standard sample A were 41.4%(w/w), 22.6%(w/w) and 36.0%(w/w), respectively. The results showed that the sugar composition of these products revealed a glucomannan pattern which is usually observed for native Aloe vera gel polysaccharides. However, for products G1~G3, T2~T4, D1 and P1, the results showed a very high amount of glucose but a very low amount of mannose. The sugar composition of these products showed a typical dextran pattern, so it was suspected that these products were adulterated with commercial maltodextrin. A typical GC chromatogram is shown in Fig.13.

Qualitative TLC analysis was performed and adulteration of maltodextrin in some products could be confirmed. $R_f$ value of standard sample A treated with cellulase was 0.53 and a single spot was observed. $R_f$ value of standard sample B untreated with cellulase was almost 0. Products A1~A9 were supposed to be treated with cellulase. $R_f$ values of maltodextrin were, however, distributed in the range of 0~0.47 and the presence of maltodextrin in the adulterated aloe gel powder could, therefore, be detected, as is illustrated in fig. 14.

The contents of maltodextrin were determined by HPLC using refractive index detector. For standard samples A and B, no peaks were observed after 5 min, as shown in fig. 15-A, B. However, peaks of maltodextrin were observed in the range 5 - 20 min and the chromatogram showed a specific elution pattern as shown in fig. 15-C. Therefore, identification and determination of maltodextrin could be performed. The standard curve for maltodextrin gave a linear response with correlation coefficient of 0.999. Consequently, in the case of products G1~G3, T2~T4, D1 and P1, the presence of dextran consistent with maltodextrin was definitive (table 6-f). Polysaccharides of products K1, A1~A9, T1, F1 and M1 were mostly consistent with Aloe vera and there was no maltodextrin contamination.

In conclusion, adulteration with commercial polysaccharides was detectable by sugar analysis and determination of ash contents. In particular, maltodextrin could be detected and the level determined by HPLC and it is therefore suggested that this method be applied to the quality control of various aloe gel powder products.

**Table 6.** Isolation and characterization of polysaccharides in various commercial aloe gel powders

| List of samples[a] | Ash contents (mg)[b] | Mass of alcohol precipitate (mg)[c] | Hexose in alcohol precipitate (mg)[d] | Mass retained upon dialysis (mg)[e] | Contents of maltodextrin (w/w, %)[f] | Relative contents (%)[g] | | | Mannose/ Glucose[h] | Adulteration with maltodextrin |
|---|---|---|---|---|---|---|---|---|---|---|
| | | | | | | Mannose | Galactose | Glucose | | |
| A[1] | 115.90 | 231.80 | 22.50 | 18.60 | N.D. | 41.41 | 22.57 | 36.03 | 1.15 | ○ |
| B[2] | 93.80 | 322.40 | 53.40 | 197.30 | N.D. | 57.83 | 22.03 | 20.14 | 2.87 | ○ |
| Maltodextrin | 0.00 | 330.70 | 223.00 | 145.50 | * | 0.00 | 0.00 | 100.00 | 0.00 | * |
| K1 | 82.80 | 320.00 | 96.10 | 163.80 | N.D. | 55.64 | 22.75 | 21.61 | 2.57 | ○ |
| A1 | 126.70 | 271.10 | 18.50 | 14.90 | N.D. | 54.53 | 29.57 | 15.90 | 3.43 | ○ |
| A2 | 188.70 | 333.30 | 10.10 | 6.20 | N.D. | 43.13 | 23.71 | 33.16 | 1.30 | ○ |
| A3 | 185.40 | 343.00 | 10.20 | 7.50 | N.D. | 42.51 | 22.03 | 35.47 | 1.20 | ○ |
| A4 | 126.50 | 334.40 | 9.10 | 7.00 | N.D. | 41.68 | 22.75 | 35.57 | 1.17 | ○ |
| A5 | 158.30 | 304.50 | 11.10 | 5.70 | N.D. | 52.30 | 20.96 | 26.74 | 1.96 | ○ |
| A6 | 167.80 | 315.30 | 13.90 | 6.60 | N.D. | 57.47 | 20.12 | 22.41 | 2.56 | ○ |
| A7 | 162.30 | 313.90 | 9.60 | 5.90 | N.D. | 39.03 | 17.66 | 43.31 | 0.90 | ○ |
| A8 | 188.80 | 347.40 | 6.70 | 3.10 | N.D. | 42.16 | 0.00 | 57.84 | 0.73 | ○ |
| A9 | 170.10 | 185.30 | 10.30 | 1.50 | N.D. | 43.25 | 0.00 | 56.75 | 0.76 | ○ |
| G1 | 0.00 | 357.80 | 285.40 | 176.50 | 57.10 | 18.65 | 0.00 | 81.35 | 0.23 | △ |
| G2 | 39.80 | 352.20 | 196.60 | 129.00 | 94.60 | 0.00 | 0.00 | 100.00 | 0.00 | × |
| G3 | 123.40 | 337.70 | 114.80 | 103.70 | 45.00 | 19.74 | 0.00 | 80.26 | 0.25 | △ |
| T1 | 148.20 | 272.40 | 38.70 | 45.50 | N.D. | 57.03 | 24.79 | 18.19 | 3.14 | ○ |
| T2 | 33.20 | 359.50 | 246.90 | 190.60 | 53.00 | 20.50 | 0.00 | 79.50 | 0.26 | △ |
| T3 | 22.80 | 369.70 | 250.40 | 141.20 | 57.80 | 24.12 | 0.00 | 75.88 | 0.32 | △ |
| T4 | 0.00 | 372.00 | 287.10 | 193.80 | 56.90 | 16.71 | 0.00 | 83.29 | 0.20 | △ |
| F1 | 130.70 | 108.90 | 34.30 | 140.30 | N.D. | 63.51 | 26.20 | 10.29 | 6.17 | ○ |
| M1 | 135.80 | 152.30 | 33.80 | 109.10 | N.D. | 64.24 | 25.09 | 10.68 | 6.02 | ○ |
| D1 | 10.00 | 383.60 | 304.90 | 170.70 | 52.50 | 0.00 | 0.00 | 100.00 | 0.00 | △ |
| P1 | 155.50 | 354.60 | 111.60 | 90.50 | 46.60 | 18.36 | 0.00 | 81.64 | 0.22 | △ |

[1), 2)]; standard samples A and B, respectively

[a]; commercial aloe gel powder except A, B and maltodextrin

[b~e]; weight (mg) in 500mg of aloe gel powder

[f]; weight of maltodextrin (mg) / weight of aloe gel powder (mg)

[g]; relative sugar contents in non-dialyzable polysaccharides isolated from aloe gel powder

[h]; content of mannose / content of glucose

○; free from maltodextrin contamination

△; adulterated with maltodextrin partially

×; mostly consisted of maltodextrin

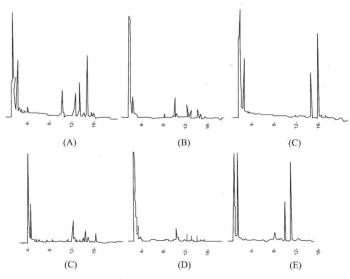

**Fig. 13.** GC chromatograms of (A) standard sample A, (B) standard sample B, (C) malto-dextrin, (D) A9, (E) K1 and (F) G2. Mannose, galactose and glucose were determined by gas chromatography after trimethylsilylation with trimethylchlorosilane hexamethyl-disilazane. 1, 2 and 3 indicate the peaks of mannose, galactose and glucose, respectively. Column, J&W DB-1 (0.25μm, 0.25mm×30m); column temperature, 160→250.; carrier gas, He 2.84mL/min; detection, flame ionization detector.

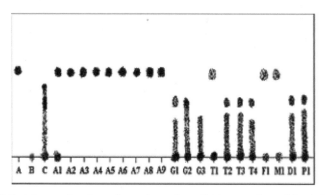

**Fig. 14.** TLC patterns of various aloe gel powders. Stationary phase, Merck Kieselgel 60 F254 precoated TLC plate; mobile phase, acetonitrile/ethyl acetate/iso-propyl alcohol/DI water = 85/20 /50/50.

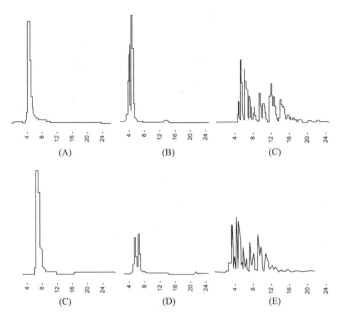

**Fig. 15.** HPLC chromatograms of (A) standard sample A, (B) standard sample B, (C) mal-todextrin, (D) A9, (E) K1 and (F) G2. Each aloe gel powder was simply dissolved into DI water, followed by injection directly onto HPLC. Column, GL Science Inertsil ODS-2 (5μm, 4.6mm × 300mm); mobile phases, 100% DI water; detector, refractive index detector.

## 5.6. Conclusion

Quality control methods for aloe and aloe preparations, which are widely used as ingredients of health foods and cosmetics, were developed. Thirteen phenolic components in Aloe barbadensis and Aloe arborescens were separated and quantified by HPLC. Among them, a quantitative HPLC method for determination of aloesin was developed. For the pharmacokinetic study of aloesin, a new and sensitive HPLC method was developed, as was a new method to detect the adulteration of commercial aloe gel powders.

**References**
1. Biltz JJ, Smith JW, Gerard JR (1963) *Aloe vera* gel in peptic ulcer theraphy: preliminary report. *J Am Osteop Assoc* 62. 731
2. Chakkodabylu SR (1989) Sensitive method for the analysis of phospholipid subclasses and molecular species as 1-anthroylnitile derivatives of their diglycerides. *JChromatogr* 491, 37-48.
3. Conner J, Gray A, Reynolds T, Waterman PG (1989) Anthrone and chromone derivatives in the exudate of *Aloe rabaiensis Phytochemistry* 28, 3551-3553
4. Conner JM, Gray AI, Waterman PG, Reynolds T (1990a). Novel anthrone-anthraquinone dimer from *Aloe elgonica J Nat Prod* 53(5),1362-1364
5. Conner JM, Gray AI, Reynolds T, Waterman PG (1990b) Anthrone and chromone components of Aloe cremnophila and A. jacksonii leaf exudates. *Phytochemistry* 29(3), 941-944
6. Crewe JEMD (1985) Aloe in the Treatment of Burns and Scalds *Minne sota Medicine* 22. 538
7. Fairbairn JW, Simic S (1960) Vegetable Purgatives containing anthracene derivatives. Part XI. Further work on the aloin-like substance of *Rhamnus purshiana* DC *J Pharm Pharmacol* 12 45T-51T
8. Franz G, Gruen M (1983) Chemistry, Occurrence and Biosynthesis of C-glycosyl Compounds in Plants, *Planta Med* 47 131-140
9. Gowda DC, Heeiisiddaiah B, Anjaneyalu YU (1979) Structural studies of polysaccharides from *Aloe vera Carbohydrate Res* 72, 201-205
10. Goto J, Goto N, Shamsa F, Saito M, Komatzu S, Suzaki K, Nambara T (1983a) New sensitive derivatization of hydroxysteroids for high-performance liquid chromatography with fluorescence detection. *Anal Chim Acta* 147, 397-400
11. Goto J, Saito M, Chikai T, Goto N, Nambara T (1983b) Determination of serum bile acids by high-performance liquid chromatography with fluorescence labelling. *J Chromatogr* 276, 289-300
12. Goto J, Shamsa F, Nambara T (1983c) Determination of 6β-hydroxycortisol in urine by high-performance liquid chromatography with flurescence detection. *J Liq Chromatogr* 6(11), 1977-1985
13. Graf E, Alexa M (1980) Stability of diastereomeric aloins A and B and their main decomposition product 4-hydroxyaloin *Planta Med* 38, 121-127
14. Hans WR, Anette B (1993)   Thin layer chromatographic screening and high performance liquid chromatographic determination of 5-hydroxyaloin A in the genus aloe. *Z Naturforsch* 8c, 1
15. Hayness LJ, Holdsworth DK (1970) C-Glycosyl compounds. part. aloesin, a C-glycosyl-chromone from Aloe sp *J Chem Soc* (C) 74, 2581
16. Haynes LJ, Holdsworth DK, Russell R (1970). C-glycosyl compounds. VI. Aloesin, a C-glycosylchromone from Aloe sp *J Chem Soc* C, (18), 2581-2586
17. Heggers JP, Kucukcelebi A, Stabenau CJ, Ko F, Broemeling, Lyle D, Robson, Martin C, Winters, Wendell D (1995) Woundhealing effects of Aloe gel and other topical antibacterial agents on rat skin. *Phytother Res* 9(6), 455-457

18. Heggers JP, Kucukcelebi A, Stabenau Catherine J, Ko F, Broemeling, Lyle D, Robson, Martuin, C, Winters and Wendell D. (1995) Wound healing effects of Aloe gel and other topical antibacterial agents on rat skin. *Phytother Res* 9, 455-457

19. Hikino H, Takahashi M, Murakami M, Konno C, Mirin Y, Karikura, M.,Hayashi T (1986) Isolation and hypoglycemic activity of arborans A and B, glycans of *Aloe arborescens var. natalensis* leaves. *Int J Crude Drug Res* 24(4), 183-186

20. Hirata T, Kushi Y, Suga T, Christensen A (1976) Structural studies of aloenin; the crystal structure of its aglycone. *Chem Lett* (4), 393-396

21. Hirata T, Suga T (1977) Biologically active constituents of leaves and roots of *Aloe arborescens* var. *nataliensis*. *Zeitchirft fur Naturforschung*, 32, 731-734

22. Holdsworth DK (1971) Chromones in Aloe species. Part I, Aloesin a C-glucosyl-7-hydroxychromone. *Planta Medica* 19, 322-325

23. Hodge JE, Hofreiter BT (1962) Determination of reducing sugar carbohydrates.Methods in *Carbohydrate Chemistry* 1. 380-394,

24. Ishii Y, Tanizawa H, Takino Y (1984) Flurophotometry of barbaloin in Aloe, *Chem Pharm Bull* 32, 4946-4950

25. Kennedy JF,   White CA (1983) in Bioactive Carbohydratesin Chemistry, Biochemistry and Biology. Ellis Horwood Limited., *Chichester* 161

26. Kim KH, Kim HJ, Park JH, Shin YG (1996) Determination of aloesin in aloe preparations by HPLC, *Yakhak Hoeji* 40, 177-182

27. Lee LM, Haggers JP, Robson MC., Hagstrom WJ. (1980) The therapeutic efficacy of Aloe vera cream in thermal injuries : Two case report. *J. Am. Anim.Hosp Assoc* 16. 768

28. Makino K, Yagi A, Nishioka I (1974) Constituents of *Aloe arborescens var natalensis*. III. Structures of two new aloesin ester. *Chem Pham Bull* 22, 1565-1570

29. Mandel G, Das A (1980) Structure of the D-galactan isolated, from *Aloe barbadensis* Miller. *Carbohydrate Res* 86, 247-257

30. Mandel G, Das A (1980) Structure of the glucoumannan isolated from the leaves of *Aloe barbadensis* Miller. *Carbohydrate Res* 87, 249-256

31. Mebe PP (1987).  2'-p-Methoxycoumaroylaloeresin, a C-glucoside from Aloe excelsa. *Phytochemistry* 26(9), 2646-2647

32. Metori A, Ogamo A, Nakagawa Y (1993)   Quantitation of monohydroxy fatty acids by high-performance liquid chromatography with fluorescence detection. *J. Chromatogr* 622, 147-151

33. Morton J (1961). Folk uses and commercial exploitation of Aloe leaf pulp. *Economic Botany* 15, 311-319

34. Nakamura H, Kan T, Kishimoto K, Ikeda k, Amemiya T, Ito K,Watanabe Y (1989) Gas chromatographic and mass spectrometric determination of aloe components in skin-care cosmetics, *Eisei Kagaku* 35, 219-225

35. Nakamura H, Okuyuma T (1990) Gas chromatographic and mass spectral determination of aloenin in skin-care cosmetics, *J Chromatogr* 509, 377-382

36. Obata M, Ito S, Beppu H, Fujita K, Nagatsu T (1993). Mechanisim of anti-inflammatory and antithermal burn action of CPase from Aloe arborescens natalensis in rats and mice. *Phytother Res* 7 (Spec. Issue, Proceedings of the International Congress of Phyto-therapy, 1991). S30-S33

37. Okamura N, Hine N, Harada S, Fujioka T, Mihashi K, Yagi A (1996) Three chromone components form *Aloe vera* leaves *Phytochemistry* 43, 495-498

38. Park MK, Park JH, Kim KH, Shin YG, Myoung KM, Lee JH (1995) Chemical constituents of *Aloe capensis Kor J Pharmacog*, 26, 244-247

39. Park MK, Park JH, Shin YG, Kim WY, Lee JH, Kim KH (1996) Neoaloesin A : A new C-glucofuranosylchromone from *Aloe barbadensis Planta Med* 62, 363-365

40. Pelley RP (1992) Aloe quality control: Current status of high pressure liquid chromatography in the quality control of Aloe barbadensis extracts. *Aloe Today* Autumm 19-26

41. Pelley RP, Wang YT, Waller TA (1993) Current status of quality control of *Aloe barbadensis* extracts. *SOFW Journal* April 255-268

42. Rai PP, Turner TD (1975) A method for the estimation of anthraquinones using densitometric thin layer chromatography. *J Chromatography* 104, 196

43. Rauwald HW, Beil A (1993a) High-performance liquid chromatographic separation and determination of diastereomeric anthrone-C-glucosyls in Cape aloe *J Chromatogr* 639, 359-362

44. Rauwald HW, Beil A (1993b) 5-Hydroxyaloin A in the genus Aloe. Thin layer chromatographic screening and high performance liquid chromatographic determination*, Z Naturforsch C: Biosci* 48, 1-4

45. Reynolds GW (1950) The Aloes of South Africa, Johan-nesberg, South Africa : *The aloes of South Africa Book Fund,*

46. Reynolds GW (1996)   The Aloes of Tropical Africa and Madagascar, Mbabane, Swaziland: *The Aloes Book Fund*

47. Reynolds T (1985) The compounds in Aloe leaf exudates: a review. *Botanical Journal of Linnean Society,* 90, 157-177

48. Saito H, Imanishi K, Okabe S (1989) Effects of aloe extract, aloctin A, on gastric secretion and on experimental gastric lesions in rats. Yakugaku Zasshi, 109(5), 335-339

49. (1991) Science and Technical Committee of the IASC, Official Certification Program For *Aloe Vera* International Aloe Science Council Inc 1-22

50. Speranza G, Dada G, Lunazzi L, Gramatica P, Manitto P (1986a) Studies on Aloe. Part 3. A C-glucosylated 5-methylchromone form Kenya aloe *Phytochemistry* 25, 2219-2222

51. Speranza G, Dada G, Lunazzi L, Gramatica P, Manitto P (1986b) Studies on Aloe. Part 4. Aloenin B, a new diglucosylated 6-phenyl-2-pyrone form Kenya aloe, *J. Nat. Prod.*, 49, 800-805

52. Suzuki Morita T, Haneda M, Ochi K, Shiba M (1986) Determination by high-performance liquid chromatography and identification of barbaloin in aloe, *Iyakuhin Kenkyu* 17, 984-990

53. Waller TA, Strickland FM, Pelley RP (1994) Quality control and biolobical activity of *Aloe Barbadensis* extracts useful in the cosmetic industry *CTMW* 64-80

54. Wawrznowicz T, Hajnos MW, Mulak-Banaszek K (1994) Isolation of aloine and aloeemodine from Aloe (Liliaceae) by micropreparative TLC *J Planar Chromatogr-Mod TLC* 7, 315-317

55. Whister RL, Wolfrom ML, BeMiller JN, Shafizadeh F (1962) *Methods in Carbohydrate Chemistry* Volume 1

56. Yagi A, Makino K, Nishioka I (1977) Studies on the constituents of Aloe saponaria Haw. III. The structures of phenol glucosides. *Chem Pharm Bull* 25(7), 1771-1776

57. Yagi A, Harada N, Shimomura K, Nishioka I (1987) Bradykinin-degrading glycoprotein in *Aloe arborescens natalensis Planta Medica* 53(1), 19-21

58. Yamamoto M, Ishikawa M, Masui T (1985) High performance liquid chromatographic determination of barbaloin in aloe *Bull Shizuoka Pref Inst Publ Hlth and Environ Sci* 28, 35

59. Yamamoto M, Ishikawa M, Masui T, Nakazawa H, Kabasawa Y (1985) Liquid chromatographic determination of barbaloin(Aloin) in foods *J Assoc Off AnalChem* 68, 493

60. Yamamoto M, Masui T, Sugiyama K, Yokota M, Nakagomi K, Nakazawa H (1991) Anti-inflammatory active constituents of *Aloe arborescens Miller Agri BiolChem* 55(6), 1627-1629

61. Yaun A, Kang S, Tan L, Raun B, Fan Y (1991) Isolation and identification of aloesin from the leaves of *Aloe vera L var chinensis(Haw.) Berger Zhongguo Ihongyao Zazhi* 16(5), 292-293

62. Zonta F, Bogoni P, Masotti P, Micali G (1995) High-performance Liquid Chromatographic profiles of aloe constituents and determination of aloin in beverages, with reference to the EEC regulation for flavoring substances *J Chromatogr A* 718, 99-106

# 6. Perspective of industrial application of Aloe vera

Park, Young In, Ph.D.[1] and Jo, Tae Hyung[2]

## 6.1. The use of Aloe for medication

Aloe, which originated from Africa under dry and hot weather conditions, is currently growing worldwide in tropical and subtropical areas. It is well recognized for its extremely high utilization qualities and its ability to adapt to various environments, making it easy for cultivation in other parts of the world with hot and humid climates. Currently there are over 360 known species of aloe and it has been used for medical purposes for probably at least 4000 years.

The use of this plant for medical needs has already been revealed by historical records in regions such as ancient Persia and Egypt, Greece, Rome, India, Africa, China, Korea and Japan. Later, with the Spanish discovery of the Americas, aloe completed its spread throughout all inhabited continents of the world.

The first record of human use of aloe is the Sumerian hieroglyphic alphabets engraved on clay tablets during the Mesopotamia civilization circa 2200 BC. According to them the aloe leaves were used as a laxative. Scriptures on the Ebers papyrus that were found inside the tombs of mummies in the Egyptian Tebe region show 12 formulas that used aloe with other ingredients to produce internal medicine or a medicine for external use, further proving the use of the plant. Material Medica, which was written by Hippocrates circa 450 BC, did not have any direct indication of aloe, but according to the Indian medication writing called the Copra (which was written during the same era) aloe was used for healing external wounds and constipation. The Greek Herbal written by Dioscoreded circa 41 AD recorded detailed information on the use of aloe vera and its extracts for wound healing, treatment of gum diseases and constipation during the Roman expedition. The hemostatic use was also reported when in dried power form.

Meanwhile, according to the Chinese 'Geboboncho' written during the Song dynasty, aloe was used for curing skin diseases named as 'Nowhae'. As this word was used later in other parts of China, Korea and Japan, it could be assumed that these 3 countries used imported aloe in the form of dried concentrate for their medical needs. In Korea, the 'DongeuiBogam' written by Joon Hur has the term 'Nowhae' added to its text as well. Aloe was cultivated in the Barbados and Curacao Islands in the Caribbean region by Spain and the Netherlands and was sold commercially to various parts of Europe during the 17th century.

Since its first recording in pharmacopeia in Germany during the 12th century, aloe has been listed as a laxative in pharmacopeia in 20 countries. Afterwards, following detailed research on the qualities and effectiveness of aloe, it has been used for a wide variety of medical purposes.

---

[1] School of Life Sciences and Biotechnology, Korea University
[2] Unigen Inc.

Among the various aloe species used for medical purposes are Socotra aloe (*Aloe perryi* Baker), which can be found in the African Somalia and Socotra Island region, Cape aloe (*Aloe ferox* Miller, *Aloe africana* Miller, *Aloe spicata* Baker), which is found near the African Cape Town region, and Curacao aloe (*Aloe barbadensis* Miller), which is found in parts of Mexico, Central America, the Caribbean and most recently Florida and southern Texas. Curacao aloe is better known as *Aloe vera*. These aloe species are currently listed in the pharmacopeia of many countries in forms of plain aloe, extract and/or powder.

The reason why extracts from aloe are used in a form of heat-treated dried powder for medical purposes is because its high water content later causes putrefaction. Before the development of technology for its processing and manufacturing, the only way to produce aloe was presumably the traditional heat dry method. For medical purposes aloe has been used as a laxative, a tonic medicine, or a peptic. Anthraquinones such as aloin and barbaloin are known to be major components.

**Aloes for medical use**

| Pharmacopeia | Species | Usage | Major Component |
|---|---|---|---|
| JP, BP, EP, etc. | Aloe ferox Miller Aloe africana Miller Aloe spicata Baker, and their Hybrids | Laxative | Anthraquinones |
| USP | Aloe barbadensis Miller (Aloe vera) | Laxative | Anthraquinones |

## 6.2. The use of aloe for food and cosmetic purposes

*Aloe vera* gel extracts began to be used for health foods/beverages and moisturizing cosmetics, starting in the United States and parts of Europe, during the 1970s. *Aloe vera* was especially cultivated and supplied in mass in the United States from its vast farming lands and automated farming methods. Aloe was later produced in the forms of gel extract and powder in order to increase its value as a product in the commercial market, contributing to increased consumption and demand for aloe.

The leaves of *Aloe vera* are thick and full of pulpy substances compared to other aloe species. The "+aloe vera gel solution"+ is a pulpy substance containing polysaccharides which is made of inner gel without the outer layer of leaves and contains almost no anthraquinones. This solution is also produced in powder form as "+aloe vera gel powder"+. This power is originally used only as a vehicle or a diluent in medicines, and as a skin moisturing agent for creams, lotions, soaps, conditioners and shampoos, whereas the yellow sap containing anthraquinones mostly between the outer layer and the gel layer was used in medical purposes.

The development of aloe as a raw material for health food is based on safety rather than pharmacological activity so that aloe has been processed to eliminate anthraquinones which show an alleviation activity for constipation.

The gel contains mainly macromolecules such as cellulose, polysaccharides and glycoproteins. The pattern of food consumption has been changing with increased intake of animal proteins and fats rather than plant celluloses. In the light of this, aloe serves as a good low calorie supplement for providing celluloses, improving fatness, and lowering cholesterol levels. Moreover, *Aloe vera* gel extracts have also been used for mucosal regeneration of the stomach and intestine and for skin care.

Other species of aloe that were expanded for use in health food purposes include *Aloe arborescens* Miller and *Aloe saponaria* Miller, but their leaves are thinner and contain much less gel. This lead to the use of whole leaves, but their use remains very limited due to concerns over the amount of gel and the side-effects of anthraquinones. However, it has been known that *Aloe saponaria* contains no anthraquinones. Three species of aloe, *Aloe vera*, *Aloe arborescens* and *Aloe saponaria,* are currently listed in the Food Code and Food Supplement Code of Korea and Japan.

**Aloes used for health food and cosmetics**

| Species | Major Component | Usage |
|---|---|---|
| Aloe barbadensis Miller (Aloe vera) | Gel Extract, Polysaccharides, Glycoproteins, and Saponins | Raw Materials for Functional Foods and Cosmetics. |
| Aloe arborescens Miller | Processed Whole Leaves Anthraquinones and Polysaccharides | Health Food Peptic and Laxative |
| Aloe saponaria Miller | Processed Whole Leaves Polysaccharides and Glycoproteins | Raw Materials for Functional Foods and Cosmetics. |

## 6.3. Modern Application and Industrialization of Aloe

Aloe has been variously used for civilian and traditional medical purposes worldwide for thousands of years, and today it remains in use as a self-medication in western and under-developed countries. As such, aloe's possession of various functions has been clinically proven. It is truly one of the very few plants that are used in such a profusion of applications.

According to the Madis Laboratories (1984), aloe has been used as a civilian traditional remedy for arthritis, gout, acne, wound healing, skin infection, headache, hypertension, digestive disorder, depilation, rheumatism, stomach ulcer, gum disease, pruritus, psoriasis, and skin burning. Aloe vera gel is being used in western countries for skin diseases such as externally on burns, and internally as a tonic (Taylor 1980). It is also being used as a moisturizing agent for cosmetics such as shaving creams and lotions, sun tanning creams and sunscreens.

The major ingredients present in the yellow sap and the gel are quite different from each other. Therefore, their clinical applications are distinct due to the difference in their pharmacological effects.

*Aloe vera* gel extracts were recognized as a cosmetic ingredient by the CTFA (Cosmetic, Fragrance and Toiletry Association) in 1989. Anthraquinones present in the yellow sap have been reported to cause skin irritation and allergic responses (Morrow *et al.* 1980) which limit the use of the yellow sap as a laxative in pharmacopeia but the gel is widely used for OTC pharmaceuticals, health foods and cosmetics.

Modern application of aloe in clinical treatment started in the 1930s. After the curing effects of *Aloe vera* on skin ulcers caused by X-ray irradiation were first proven by modern scientific methods (Collins and Collins 1935), many studies have been conducted for the treatment of various skin diseases such as ulcer, inflammation, and burning caused by X-ray irradiation. With the development of nuclear power, the United States government conducted research on the curing capabilities of *Aloe vera* on heat and radiation burns in order to introduce its use in the military (Ashley *et al.* 1957). In addition the research on *Aloe vera* was accelerated in 1952 when a Japanese boat was contaminated by passing through the US Bikini Islands during a hydrogen bomb test. The US government procured medication using *Aloe vera* to treat the radiation burns that the sailors suffered from the irradiated dust. After this incident, scientists became interested in research on the ingredients of aloe which expressed those effects other than anthraquinones as a laxative.

After aloe ointment was recognized as an OTC medication for healing wounds on the skin by the USFDA in 1959, extensive research on clinical properties was conducted in terms of its pharmacological and treatment effects. Aside from its ability to heal skin wounds, *Aloe vera* has also been proven to cure oral infections and diseases related to the gums (Bovik 1966; Payne 1970). In addition, it has also been verified that *Aloe vera* can be clinically applied to chronic ulcers on the foot (El Zawahry *et al.* 1973), can stimulate growth of normal human cells by its lectin (Winters *et al.* 1981), and can cure skin burns and even frostbite (Lorengetti *et al.* 1964; Bruce 1967).

There are many reports in regards to the anti-inflammation effects, such as anti-bradykinin activity of a glycoprotein in *Aloe saponaria* (Yagi *et al.*, 1982), the bradykininase and carboxypeptidase activities from *Aloe arborescence* extract (Fujita *et al.* 1976, 1979), the effect of aloctin A as a lectin from *Aloearborescence* on edema and adjuvant arthritis (Saito *et al.* 1982), and the suppression of prostaglandin synthesis by *Aloe vera* gel (Heggers and Robinson 1983). Moreover, Cape Aloe and aloctin A have been reported to contain anti-cancer activity (Soeda 1969; Imanish *et al.* 1981). For immune-modulation activities, there are polysaccharides and glycoproteins of *Aloe arborescence* on allergic asthma, and the recovery of immune suppression caused by UV irradiation with *Aloe vera* gel (Strickland *et al.* 1994). In addition, other reported activities include the curing effect of AIDS by acemannan (Carpenter 1992), the anti-oxidation by aloeresin and aloesin (Yu *et al.* PCT patent US 1995), and the decrease in blood glucose level (Ghanam *et al.* 1986) of *Aloe vera* gel.

Based on these findings, the production of aloe (especially *Aloe vera*) has been industrialized over the last 50 years by growing and processing in large quantities for OTC medications, health foods and cosmetics. Accordingly, mass cultivation and processing technologies have also been developed.Many technological improvements have been achieved such as the High Temperature Short Time (HTST) method for hygienic preparation of the gel, technology for the stabilization of the gel concentrate without destroying the effective components, and the improved freeze drying method or spray drying method to preserve the quality of the gel extract as much as possible.

As such, aloe as a raw material has been produced massively and made easily available. However, despite this expansion in production and various research papers, the development of more specialized and professional medications has lagged well behind, presumably because the excessive commercial mentality of company leaders. Instead of investment in R&D for new products based on the spearhead scientific technology, they have been simply relying on improvements of traditional methods to boost productivity in a strictly commercial manner. In this regard, aloe was simply treated as a cure for all injuries and diseases by these company leaders, which led to criticism from scientists and the perception that aloe is a traditional, homemade medicinal ingredient and nothing more. For example, companies advertised that aloe was used as a panacea by King Alexander or Queen Cleopatra, which simply increased the level of distrust for aloe among the potential customers.

For the sake of scientific R&D on aloe, the problems have been encountered in this area. First, aloe scientists did not initially establish proper experimental plans. Most research was limited to fulfilling basic academic knowledge and was not able to be expanded further in detail. Second, a false belief was developed that aloe is most effective when its various substances are used together, rather than as separate substances. This led to the extraction of aloe as a whole, and limited the need for any kind of in-depth research. The lack of scientific experiments made it difficult to produce patented, medical products using separate, purified aloe substances. In addition, the big pharmaceutical companies did not actively invest in investigating this research matter. Third, because aloe was considered to be a traditional household remedy, aloe was not able to attract serious scientific R&D into the potential for its sale in first level processed forms of extract or powder.

Despite these limitations, aloe has been widely used for medical purposes. Natural products extracted from plants started to be used in the late 18th century with the discovery of various organic acids. Especially, purification of morphine from opium in the early 19th century led to the use of effective substance(s) isolated from plants, rather than the whole plants themselves.

After this, a lot of natural products were isolated and identified from various plants, from which many organic compounds were designed and synthesized as lead compounds for improving physiological activity and decreasing toxicity. In many cases, although natural products could be disregarded because of their relatively weak physiological activities, it must be considered that their pharmacological activities are sometimes very useful for causal treatment of chronic diseases.

From this background, the Namyang Aloe Corporation formed the CAP research and planning committee (later known as the CAP project) in 1993, with Professor S.K. Lee of Seoul National University as the chairman, in order to instigate R&D on aloe. Based on a review of various existing research materials regarding aloe, several strategic principles were established.

First, the objectives of the CAP Project were defined to develop technologies for application of properties of aloe rather than the simple generation of academic knowledge.

Second, the ultimate goal of the CAP Project was set as the development of new drugs. During this process, it was decided that scientific knowledge and information obtained would be applied for the development of new products of health foods or functional cosmetics. Therefore, the project was empowered to draw an R&D road map emphasizing a collection of scientific findings and, furthermore, to establish a continuous investment plan to form an economic cycle for the project.

Third, the project was designed to target designated effects that have been recognized for centuries. Based on these target effects, a methodology for the isolation and assay was established.

Fourth, the CAP Project pursued a systematic and well-designed R&D program. The participating scientists were divided into two groups ; substance isolation and analysis team and an assay team (Ed- no respective comparison). Once, a certain isolated substance is matched with a certain function, scientists related with the activities will form another research team to cooperatively develop a new product.

Fifth, the CAP Project was designed for causal treatment rather than symptomatic treatment of diseases. So far, many natural substances expressing strong pharmacological activities have been isolated and used for the treatment of diseases. In addition, many drugs are effective in the treatment of acute diseases but only show short term effects and are not effective in the treatment of chronic diseases. In this regard, interest in the application of natural products has been expanded due to their less effective but sustained activity. Therefore, the project was focused on the elucidation of the pathology of chronic diseases, and on the identification of physiologically active components for causal treatment by understanding the mechanisms of their actions.

Sixth, the CAP Project was administered strategically through the formation of a management and planning committee because cooperation among participants, including scientists and the company, is essential for fast decision making and effective project operation.

## 6.4. Application and Prospects for aloe

### 6.4.1. Application of Anthraquinones

Anthraquinones are major substances found in aloe, and are known for their laxative, anti-microbial and anti-cancer activities. Laxation activity has already been registered in many pharmacopeias from various countries, but this activity may merely be a side effect for the internal use of anti-microbial or anti-cancer agents. In addition, they can cause skin irritation. Moreover, their anti-microbial activity is not predominant over existing antibiotics on the market; therefore, they were excluded from the project.

### 6.4.2. Application of Phenolic Compounds

Phenolic compounds are the second major substances found in aloe. Aloesin has been reported to have a skin whitening effect (Yagi 1977).

The pharmacological mechanism of aloesin on the skin whitening effect has been studied based on previous reports and its activity compared with arbutin which is known to be a major substance for skin whitening cosmetics. The use of aloesin in combination with arbutin has been shown to be more effective in skin whitening, i.e., in synergistically inhibited tyrosinase, which is the key enzyme for the synthesis of melanin pigment in the skin. Moreover, aloesin has been confirmed through research to be effective in liver and kidney cell proliferation. Based on these findings, new health foods such as 'Namyang 931', 'Aloesin' or 'Lament Aloe Whitening' have been launched successfully on the market. The new function of aloesin's antioxidation activity has been also identified. This activity has been found to be a useful protector against kidney damage when anti-cancer drugs of the cisplatin family are used for cancer treatment.

### 6.4.3. Application of Glycoproteins and β-sitosterols

The wound healing effect is a major function of *Aloe vera* which has been identified repeatedly in the project by the *ex vivo* experiments such as cell proliferation assay. Angiogenetic activity has been newly identified in aloe as being stimulated by β-sitosterol and a patent was filed for medical applications. Although β-sitosterol is not the sole component of aloe, it is already marketed as an alleviation agent for either the level of blood cholesterol, prostate megaly or gum disease. The new function of β-sitosterol on wound healing is now under development in terms of blood circulation malfunction.

Furthermore, the glycopeptide isolated from aloe was shown to contain a wound healing effect by stimulating the proliferation of skin cells. Research into the purification of this material is ongoing.

Another glycoprotein, alprogen, has been shown to be very effective on allergies. It strongly inhibited the release of histamines and leukotrienes together. Its structure has been partially identified and its genes have been cloned successfully. The study of recombinant alprogen as well as natural alprogen is ongoing for the development of an anti-allergic drug.

The first stage of the CAP project was pursued from 1993 to 2001. Various new functions were generated such as the wound healing effect, angiogenetic activity, anti-allergic activity, skin whitening effect, anti-oxidant activity, cell proliferation activity on the liver and kidney, and immune modulation activity on the isolation of cognate substances. In addition, the processing of aloe as a raw material and its quality control were standardized. Many research papers were generated and many patents filed.

Such research on the development of new aloe products has led to the accumulation of considerable industrial experience of the R&D necessary to develop natural remedies and functional materials. This research has actually stimulated aloe to occupy the top market position exclusively. In fact, many benefits were accrued with the approval of aloe as a nutraceutical by the regulation law of health foods that was established in Korea in 2003.

The second stage of the CAP R&D project has been progressing under the management of the newly organized committee (with Professor Young In Park of Korea University as the committee chairman) on research based on the findings of the first stage studies such as the wound healing activity of glycopeptide, the anti-allergic activity of alprogen, the immune modulation activity of acemannan and the activity for Alzheimer disease treatment. This wide body of research will certainly contribute to increasing the market value of aloe.

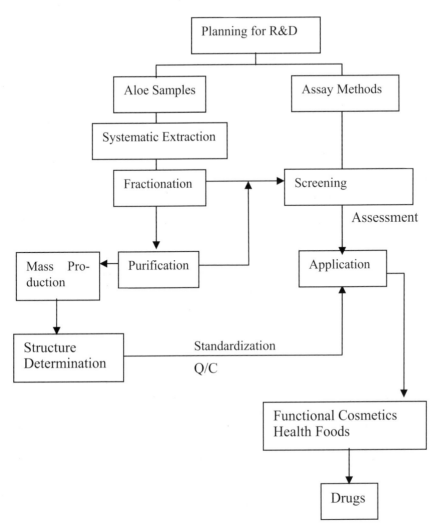

**Scheme 1.** Systematic Diagram of CAP Project R&D

# References

1. Ashleye AD (1983) Applying heat during processing of commercial Aloe vera gel Erds Int;   1(1):40-44

2. Bruce WG (1967)Investigations of antibacterial activity in the aloe. S. Afr. Med. J., 41(38):984

3. Fujita K, Teradaira R, Nagatsu T (1976) Bradykininase activity of aloe extract.Biochem Pharmacol 25(2):205

4. Fujita K, Ito S, Teradaira R, Beppu H (1979) Properties of a carboxypeptidase from aloe. Biochem Pharmacol 28(7);1261-1262

5. Ghannam N, Kingston M, Al-Meshaal IA, Tariq M, Parman NS, Woodhouse N (1986) The antidiabetic activity of aloes: Preliminary clinical and experimental observations. J. Hormone Res, 24(4);288-294

6. Heggers JP, Pelley RP, Hill DP, Stabenau J, Winters W (1992) Progressive tissue necrosis and the effects of Aloe. Abstr Gen Meet Am Soc Microbiol; Vol.92:87

7. Imanishi K, Ishiguro T, Saito H, Suzuki I (1981) Pharmacological studies on a plant lectin, Aloctin A. I. Growth inhibition of mouse methylcholanthrene -induced fibrosarcoma (Meth A) in ascites form by Aloctin A. Experientia 37(11): 1186-1187

8. Morrow DM, Rapaprot MJ, Strick RA (1980) Hypersensitivity to aloe. Arch. Dermatol., 95(11):1064-1065

9. Robson MC, Heggers JP, Hagstrom WJ Jr (1982) Myth, magic, witchcraft, or fact? Aloe vera revisited, J Burn Care Rehabil, 3(3):157-163

10. Saito H, Ishiguro T, Imanishi K, Suzuki I (1982) Pharmacological studies on a plant lectin aloctin A. II. Inhibitory effect of aloctin A on experimental models of inflammation in rats. Jpn J Pharmacol, 32(1):139-142

11. Shida T, Yagi A, Nishimura H, Nishioka I (1985) Effect of Aloe extract on peripheral phagocytosis in adult bronchial asthma. Planta Med, 3:273-275

12. Struckland FM, Pelley RP (1994) Prevention of ultraviolet radiation-induced suppression of contact and delayed hypersensitivity by aloe barbadensis gel extract. Journal of investigative Dermatology   102(2):197-204

13. Winters WD, Benavides R, Clouse WJ (1981) Effect of Aloe extracts on human normal and tumor cells in-vitro. Econ Bot, Vol.35 Nc.1; 1981:89-95

14. Yagi A, Harada N, Yamada H, Iwadare S, Nishioka I (1982) Antibradykinin active material in Aloe saponaria. J Pharm Sci 71(10):1172-4